D0773368

Common to This Country

C. 1

Common to This Country

BOTANICAL DISCOVERIES OF LEWIS AND CLARK

text by **SUSAN H. MUNGER**

illustrations by **CHARLOTTE STAUB THOMAS**

foreword by **VERLYN KLINKENBORG**

NEW YORK

Grateful acknowledgment is made to the American Philosophical Society for permission to reproduce historical documents pertaining to Lewis and Clark's expedition and to the Academy of Natural Sciences of Philadelphia, Ewell Sale Stewart Library, for images of botanical specimens.

Published by Artisan
A Division of Workman Publishing, Inc.
708 Broadway
New York, New York 10003-9555
www.artisanbooks.com

Library of Congress Cataloging-in-Publication Data
Munger, Susan H.
Common to this country : botanical discoveries of Lewis and Clark / text by Susan H. Munger ; illustrations by Charlotte Staub Thomas ; foreword by Verlyn Klinkenborg.
p. cm.
Includes bibliographical references and index.
ISBN 1-57965-224-7
1. Botany—West (U.S.) 2. Botany—West (U.S.)—Pictorial works. 3. Lewis and Clark Expedition (1804-1806) 4. Plant collecting—West (U.S.)—History.
I. Thomas, Charlotte Staub. II. Title.

QK133.M86 2003
581.973—dc21 2003052302

Printed in Italy
10 9 8 7 6 5 4 3 2 1

Book design by Vivian Ghazarian

To the Tennis Group

CONTENTS

FOREWORD "There is not a sprig of grass that shoots uninteresting to me," wrote Thomas Jefferson. And when in June 1803 it came time to direct the attention of his explorers, Captains Meriwether Lewis and William Clark, as they prepared themselves for their great expedition, Jefferson instructed them in a detailed letter to keep a close watch on "the soil & face of the country it's growth & vegetable productions, especially those not of the U.S." Reading Jefferson's letter is a bracing reminder of the intellectual omnivorousness he expected these two remarkable men to show. It was not enough that they find their way across the Louisiana Purchase, to the shores of the Pacific Ocean, and back again. Their mission was to notice and record literally everything about the world they passed through—weather, geography, latitude, longitude, cultures, languages, traditions, laws, commerce, animals, minerals, fossils, geology—and to collect whatever examples they could. Lewis and Clark acquitted themselves unusually well when it came to the botany of this new world.

The Corps of Discovery's expedition will always be the central narrative of American expansion, as long, that is, as we allow ourselves to feel its imaginative power. Had it been an expedition of conquest, merely a way of notarizing the acquisition of new territory, its place in our memories might have dwindled over the years. But because it was above all a scientific expedition, and because Lewis and Clark were so vigilantly *aware,* the journal of their expedition allows us to witness, as often as we choose to read it, the very invention of the American west. In a real sense, the language of the scientific tasks that Jefferson had sketched out for them, their roles as collectors and journal keepers, gave them the eyes they needed to truly see what was before their eyes.

Unlike most of history's great plant collectors, Lewis and Clark had much, much more on their minds than collecting plants. Still, Lewis amassed a collection of some two hundred specimens, most of which still survive, and in his journal he kept a running commentary on the wonders of the West. He noted beauty—and who wouldn't, seeing so much camas in bloom on the prairie that it resembled "a lake of fine clear water"?—and utility as well, sometimes economic, sometimes medicinal, and sometimes the ornamental utility of plants that might be serviceable in gardens in the East. In his commentary we can hear the son of a noted herbalist as well as the chosen observer of a president who was a committed gardener and farmer. As Lewis tells what he has found, we can almost see him transplanting the specimen in his mind, deciding how it might be used back home.

Many of the plants Lewis described and collected, including the ones featured here, are now so familiar to us that it is hard to imagine them being newly discovered. But that's the way it is with the familiar. Our gardens abound with plants that have become tame through our exposure to them, so that they evoke for us a sense of home, of place, no matter how distant or exotic their origins really were. Reading these pages, you can't help but feel a strange kind of nativism. Lewis and Clark went looking for "plants not of the U. S." when the United States was only the narrow eastern seaboard. What they found turned out to be, to us, very much the plants of the United States.

Common to This Country reminds us once again of the scientific stature of this great expedition, as well as its romantic freshness. With this book in hand, you could begin a Lewis and Clark garden, filled with plants of their discovering. But to make such a garden you would find yourself in need of wet spots and dry spots, high altitudes and low, great warmth and extreme cold, exposed sites and crevices nearly cut off from the elements. You would find yourself needing, in other words, the lay of the land on the far side of the Mississippi along a northwesterly route bound for the Oregon shores of the Pacific.

Verlyn Klinkenborg

INTRODUCTION The contribution Meriwether Lewis and William Clark made to North American botany is immense. In 1804 the explorers set out to cross the North American west and on the way discovered a whole new world of botany. They found plants native to arid plains, to rugged mountains, to cloud-covered forests. They collected, pressed, and carried home more than two hundred specimens, many then unknown to Western science. Some they described in great detail in their journals; others they did not even mention. They collected plants that reminded them of familiar species from home and others that they learned about from the native inhabitants. They kept their eye out for plants of potential commercial value, as well as ones that were simply appealing or curiously different.

Meriwether Lewis

The Corps of Discovery, as the Lewis and Clark expedition is known, was the brainchild of Thomas Jefferson, whose appetite for knowledge of the natural world made him intensely interested in the land beyond the Mississippi River, which formed the western boundary of the United States until May of 1803. The expedition would follow the Missouri River across the newly acquired Louisiana Purchase and over the Rocky Mountains to the Pacific Ocean, gathering information along the entire route. Its primary objective was to find a water passage for commerce to the Pacific Ocean. Another was to let the Native Americans and the occasional Englishman, Frenchman, or Spaniard living in this vast space know that it now belonged to the United States.

Many aspects of this expedition have amazed people. The unwavering friendship between the two captains and their unique sharing of responsibilities, the extensive collection of objects they brought back, and the fact that all but one member survived are often cited. The story of Sacagawea, the Shoshone woman who accompanied the expedition carrying her infant son all the way from North Dakota to the Pacific Coast and back, has fascinated people for over a century. William Clark's manservant, York, was the first black man to set eyes on the Pacific Ocean north of Mexico. Especially remarkable is the extraordinary legacy in the form of the journals that the two captains and several others kept throughout the entire two-and-a-half year trip.

Thomas Jefferson charged Lewis and Clark, two army officers, with a mission far broader than merely securing the new territory. He wanted them to find out everything about it. He expected the captains to make careful observations, keep detailed records, and prepare maps that would extend the scope of knowledge. An expedition setting out today with a comparable mission would need a team of a hundred experts. When it came to the plants Jefferson wanted specifics; he wanted to know "the dates at which particular plants put forth or lose flowers, or leaf," especially of those not common to the United States. Lewis was the man to do the botanizing for the expedition; Clark was the mapmaker.

Lewis was born in Virginia and spent his earliest years there and in Georgia. From a young age he was in the habit of rambling off on his own for long stretches, sometimes spending the night in the woods. As a young army officer he served on the western frontier between the Appalachian Mountains and the Mississippi River. When Jefferson became president, he took Lewis on as his personal secretary. Lewis invited William Clark, a fellow Virginian with whom he had served in the army, to cocaptain the expedition. Clark's knowledge of surveying and innate talent for mapmaking made him an ideal partner. The two spent several months studying astronomy and mapmaking in preparation for the journey. Both men seemed to sense that they would make a winning partnership.

William Clark

Jefferson sent Lewis to Philadelphia to study with various experts, including Benjamin Smith Barton, professor of botany at the University of Pennsylvania and author of the first botany text published in the United States. Lewis took Barton's text and other reference books on the expedition. Probably it was Barton who taught Lewis how to press and dry plant specimens so they could be brought back to Philadelphia and studied by experts. Some of Lewis's plant specimens would be shipped back to St. Louis after the first winter in 1805. The rest were carried across the continent. Today there are 226 of Lewis's pressed plant specimens in the Lewis and Clark Herbarium at the Academy of Natural Sciences in Philadelphia. (For more information, see page 123.)

The Louisiana territory, extending northwest from New Orleans across the plains to the Continental Divide north of Texas, was owned by Spain from 1762 until 1800 when it agreed to cede the region to France. Preparations for the expedition were under way when in April 1803 the Napoleonic government sold the territory to the United States, suddenly doubling the size of the nation. Now the expedition would travel much of the way across U.S.-owned land.

Ships from Europe and the United States had been exploring the Pacific Coast for many years, and there was some familiarity with geography of the West Coast. The big unknown was the mountains. Maps of the time showed a little strip of modest-size mountains or no mountains at all. No one had any idea how high and how extensive they are. If they had known, there might never have been such an expedition. When Lewis and Clark crossed range after range of huge peaks they realized that a water route to the Pacific did not exist. The explorers repeatedly and with awe referred to the mountains as stupendous, tremendous, and immense. The expedition brought to an end any dreams of finding a Northwest Passage.

When the Corps of Discovery set out on May 14, 1804, Meriwether Lewis was twenty-nine years old and William Clark was thirty three. They led a group of about forty-five men, which included French watermen, or *engages*, hired to help on the first stretch upriver. It was difficult work to maneuver a keelboat and two flat-bottomed pirogues against the current and wind, dodging trees floating down the river, avoiding collapsing river banks. The boats were periodically swamped and supplies and equipment had to be unloaded and set out to dry. In the late fall of 1804, the expedition reached central North Dakota. Here the group spent the winter with the Mandan Indians. The Mandans were friendly and informative, giving them much good advice about what lay ahead. During that winter Lewis and Clark invited Sacagawea and her French husband, Touissant Charbonneau, to accompany them as translators. In the spring of 1805, six men returned to St. Louis in the keelboat loaded with objects and reports of what they'd seen so far. The rest of the expedition, now numbering thirty-three, headed west across the northern plains.

Soon the explorers saw in the distance huge mountains like nothing they had anticipated. They battled daily swarms of mosquitoes and gnats, sharp-spined cactuses, stinging plants, and sharp stones and quickly learned to respect the ferocious grizzly bears. In late summer they had arrived at the Rocky Mountains. Here they abandoned their boats and, with Sacagawea as liaison, purchased horses from Shoshones, who also guided them over the Rockies. When the captains realized their size and extent, they knew they would not make it back to the United States that year and would have to spend the winter on the Pacific coast. That winter they endured almost endless days and nights of wet stormy weather. Their clothes rotted and they could not find much game for food. The following spring they left the coast and headed east, back over the mountains, across the plains, down the Missouri, arriving home in late September 1806.

Although they were written in an age when diarists did not often express emotions or personal feelings, the journals evoke wonderfully rich visions. Lewis did not suppress his delight when he wrote how beautiful the plains looked with the prickly pears in bloom or how enchantingly the birds sang after a refreshing rain. His descriptions of plants and animals are filled with precise details expressing a deep appreciation of his subjects. Years later Jefferson wrote in praise of Lewis and "his talent for observation which had led him to an accurate knolege of the plants & animals of his own country." This knowledge gave Lewis the ability to identify and describe the new plants common to the country west of the Mississippi.

Some of the plants that Lewis collected have become familiar as garden plants, such as Oregon grape holly, perennial flax, and osage orange. Many are still only to be enjoyed in their native habitat. Others are related to familiar species and hybrids, including monkeyflowers, lupines, lewisias, geums, honeysuckles, and mock oranges.

At the beginning of the nineteenth century the United States was on the verge of change. With the acquisition of the Louisiana Territory, people from the East rushed westward. Just a few years after the expedition paddled into St. Louis, Robert Fulton launched the first commercially successful steamboat on the Hudson River. The botanist Thomas Nuttall set out to collect and describe many of the same plants Lewis had collected. In 1811 John Jacob Astor mounted an expedition to follow the route of Lewis and Clark and set up a fur trading company on the Pacific coast. The longest-lived member of the expedition, Sergeant Patrick Gass, died one year after the completion of the transcontinental railroad.

When the men returned to the United States they were greeted as heroes and feted from St. Louis to Washington, D.C. Lewis's story thereafter is a sad one. Jefferson misjudged his capabilities and appointed him governor of Louisiana Territory. Lewis was not suited for political appointment and proved to be a poor administrator. His mental and physical health declined and in October 1809 he died, probably by his own hand. Clark soon headed home, married, and raised a family. He was appointed Superintendent for Indian Affairs and served for many years before his death in 1838. Clark oversaw the effort to get a much condensed version of the journals published in 1814. Then interest in the expedition seemed to dry up. As the first centennial of the expedition approached, interest revived and the first publication of the full journals appeared in 1904. Since then, more of the expedition's documents have come to light and the botanical specimens have been placed in proper housing.

The story of Lewis and Clark is one of America's great sagas. The amazing wealth of original material now available, along with the greater perspective that comes with time, makes the experiences of the Corps of Discovery more fascinating than ever.

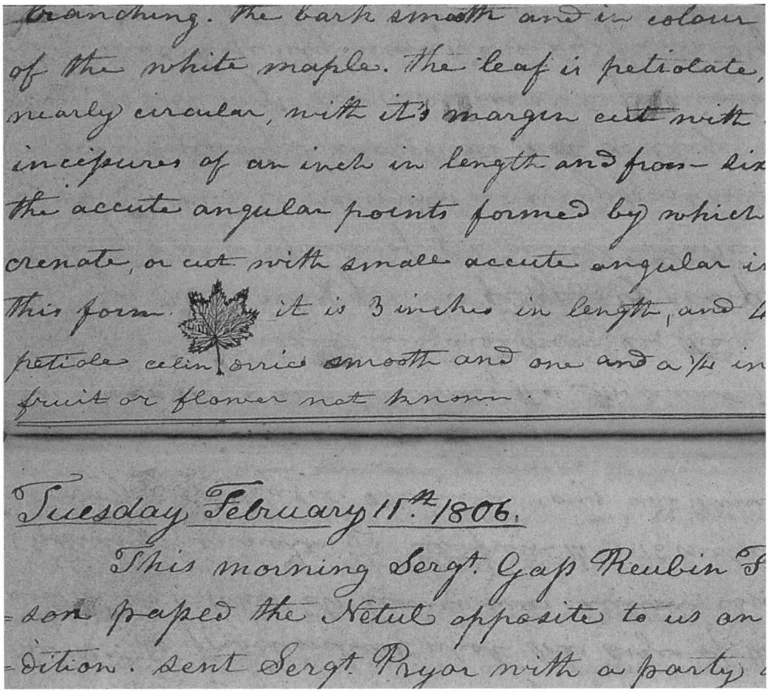

branching. the bark smooth and in colour
of the white maple. the leaf is petiolate,
nearly circular, with it's margin cut with
incisures of an inch in length and from six
the accute angular points formed by which
crenate, or cut with small accute angular i
this form. it is 3 inches in length, and 4
petiole celindric smooth and one and a 1/4 in
fruit or flower not known.

Tuesday February 11th 1806.
This morning Sergt. Gass Reubin F
son passed the Netul opposite to us an
dition. sent Sergt. Pryor with a party

LEWIS'S JOURNAL ENTRY FOR FEBRUARY 10, 1806, WITH DRAWING OF VINE MAPLE, *ACER CIRCINATUM*, A NATIVE OF THE PACIFIC NORTHWEST COAST. LEWIS COLLECTED A SPECIMEN IN THE FALL OF 1805 AND MADE THE DRAWING WHILE AT FORT CLATSOP. THE NAME REFLECTS ITS TRAILING HABIT. IT IS SOMETIMES USED AS AN ORNAMENTAL.

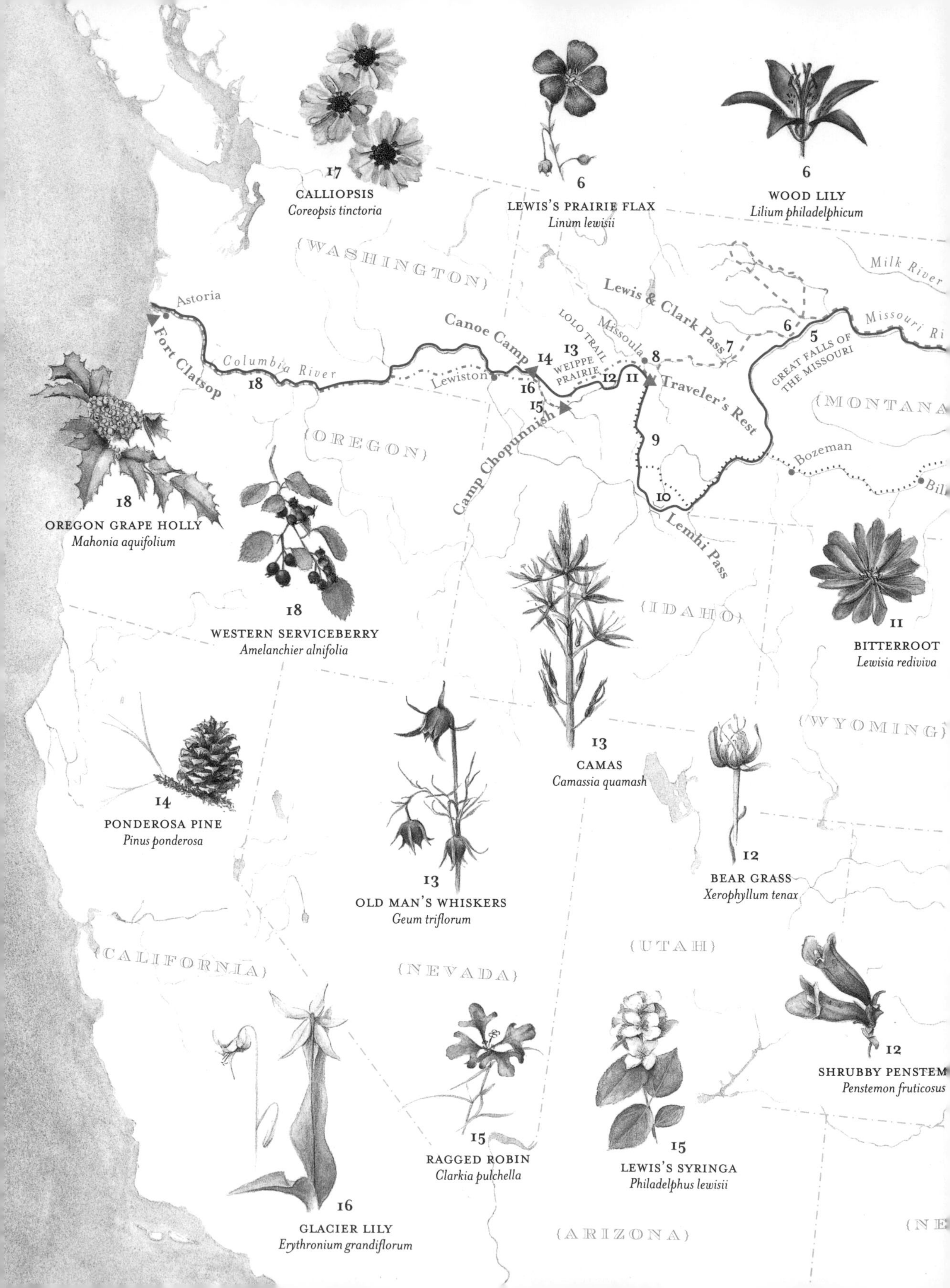
17
CALLIOPSIS
Coreopsis tinctoria
6
LEWIS'S PRAIRIE FLAX
Linum lewisii
6
WOOD LILY
Lilium philadelphicum
{WASHINGTON}
Milk River
Astoria
Fort Clatsop
Columbia River
18
Lewiston
Canoe Camp
Lewis & Clark Pass
LOLO TRAIL
Missoula
WEIPPE PRAIRIE
Missouri Ri
GREAT FALLS OF THE MISSOURI
Traveler's Rest
Camp Chopunnish
{OREGON}
{MONTANA
Bozeman
Lemhi Pass
18
OREGON GRAPE HOLLY
Mahonia aquifolium
18
WESTERN SERVICEBERRY
Amelanchier alnifolia
{IDAHO}
11
BITTERROOT
Lewisia rediviva
{WYOMING}
13
CAMAS
Camassia quamash
14
PONDEROSA PINE
Pinus ponderosa
13
OLD MAN'S WHISKERS
Geum triflorum
12
BEAR GRASS
Xerophyllum tenax
{CALIFORNIA}
{NEVADA}
{UTAH}
12
SHRUBBY PENSTEM
Penstemon fruticosus
15
RAGGED ROBIN
Clarkia pulchella
15
LEWIS'S SYRINGA
Philadelphus lewisii
16
GLACIER LILY
Erythronium grandiflorum
{ARIZONA}
{NE

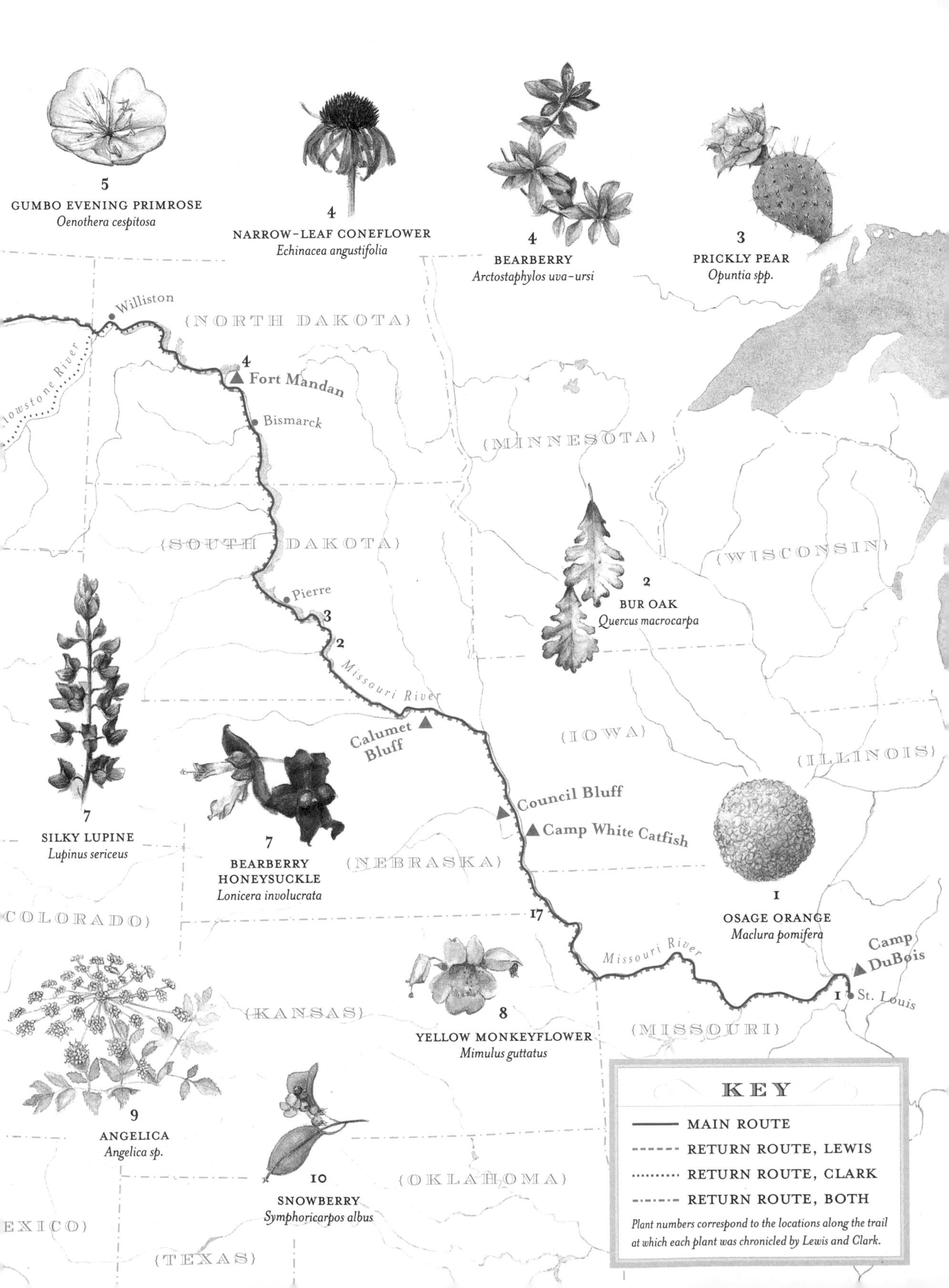

5
GUMBO EVENING PRIMROSE
Oenothera cespitosa
4
NARROW-LEAF CONEFLOWER
Echinacea angustifolia
4
BEARBERRY
Arctostaphylos uva-ursi
3
PRICKLY PEAR
Opuntia spp.
Williston
{NORTH DAKOTA}
Yellowstone River
4
Fort Mandan
Bismarck
{MINNESOTA}
{SOUTH DAKOTA}
{WISCONSIN}
Pierre
3
2
2
BUR OAK
Quercus macrocarpa
Missouri River
Calumet Bluff
{IOWA}
{ILLINOIS}
7
SILKY LUPINE
Lupinus sericeus
7
BEARBERRY HONEYSUCKLE
Lonicera involucrata
Council Bluff
Camp White Catfish
{NEBRASKA}
1
OSAGE ORANGE
Maclura pomifera
{COLORADO}
17
Missouri River
Camp DuBois
1
St. Louis
{KANSAS}
8
YELLOW MONKEYFLOWER
Mimulus guttatus
{MISSOURI}
9
ANGELICA
Angelica sp.
10
SNOWBERRY
Symphoricarpos albus
{OKLAHOMA}
{MEXICO}
{TEXAS}
KEY
MAIN ROUTE
RETURN ROUTE, LEWIS
RETURN ROUTE, CLARK
RETURN ROUTE, BOTH
Plant numbers correspond to the locations along the trail at which each plant was chronicled by Lewis and Clark.

OSAGE ORANGE
Maclura pomifera

Osage Orange

MACLURA POMIFERA

ON MARCH 26, 1804, Captain Meriwether Lewis wrote to President Thomas Jefferson, "I send you herewith inclosed, some slips of the Osages Plums, and Apples. I fear the season is too far advanced for their success." In the letter he described the osage apple, which we now call osage orange, in some detail. The United States had just acquired from France the Louisiana Purchase, a vast territory between the Mississippi River and the Rocky Mountains. Lewis and Captain William Clark were in St. Louis, recruiting members for the Corps of Discovery, purchasing tools and supplies, and building boats for their journey up the Missouri River and across the mountains to the Pacific Ocean. Pierre Chouteau, a prominent citizen of St. Louis who had lived with the Osage nation, introduced the osage orange to Captain Lewis.

Chouteau had obtained his plants from an Osage Indian village west of St. Louis and had been growing the trees for about five years when Lewis met him. This medium-size tree is native to Arkansas, Oklahoma, and Texas. Lewis wrote that the general contour was similar to hawthorn, that "its smaller branches are armed with many single, long & sharp pinated thorns," but that it was less branched, it reached thirty feet in height, and its bark was lighter in color. Because of the early season, Lewis was not able to describe the leaf or flower to Jefferson; however, he did pass on information from the Indians, who gave an extravagant account of the aroma of the mature fruit. In late summer, guided by the smell, they obtained wood from the trees, which they valued highly. It is hard, tough, and durable and makes excellent bows; another name for the tree is *bois d'arc,* bow wood. The Indians thought the fruit poisonous and did not eat it, though animals fed on it. Based on their description, Lewis wrote that the fruit was the size of a large orange, globular, and of a fine orange color (it is generally described as greenish yellow). He gave more botanical details of the fruit and concluded by saying that Chouteau's trees had not yet produced either flowers or fruit.

Although Lewis had seen only immature trees during the winter months, this letter to Jefferson demonstrates his eagerness to report on new plants of scientific and possible commercial interest. He even gave Jefferson the name of someone in St. Louis who could send more specimens to the East should his not be viable.

Later writings by Jefferson and Bernard McMahon, a Philadelphia plantsman, indicate that specimens provided by Lewis were viable. McMahon, one of several men whom Jefferson entrusted with plants and seeds brought back by the Corps of Discovery, had come to the United States from Ireland and established a thriving business as a nurseryman and seed supplier. Over the years he provided Jefferson with many plants for the gardens at Monticello and he has been described as Jefferson's gardening mentor. Nearly twenty years after Jefferson received Lewis's slips of osage, he wrote to a friend of two trees he thought quite beautiful. One was "the celebrated Bow wood of Louisiana, which may be planted in the spring where it is to stand as it bears our climate perfectly. It bears a fruit of the size and appearance of an orange, but not eatable." Osage orange was thriving in Jefferson's garden, thanks to Merriwether Lewis's having found it in the Louisiana territory eighteen years earlier.

By mid-nineteenth century osage orange had become a favorite hedge plant; the eleventh edition of McMahon's popular *American Gardener* still paid tribute to Lewis and Clark for having introduced osage orange to the country. Looking at things from the gardener's perspective, the book advised that every person having a collection of trees and shrubbery should have osage orange because of its rich and beautiful foliage. The plant was easy to grow from seed, layers, or cuttings and was valuable for hedging because it grew rapidly and the long thorns made it impenetrable. Indeed, osage orange hedges are still used to enclose livestock; hence it is also called hedge tree or hedge apple. *American Gardener* noted that osage orange could also be grown as a standard or ornamental and the branches could be trained to form a summerhouse; one garden in Philadelphia had a tree large enough for thirty people to dine under. Yet another interest in this plant, a member of mulberry family, arose from the silkworm's liking for feeding on its leaves.

Although Lewis was the first to collect and describe *Maclura pomifera,* it was named in honor of the early American geologist William Maclure. It is a tough plant, tolerating both wet and dry conditions, wind, extreme heat, and acidic and alkaline soils. These characteristics enable it to survive in its native habitat and also in cities and other difficult sites. Today there are trees growing in Philadelphia and at the University of Virginia that are said to be direct descendants of the cuttings sent back by Lewis.

When osage orange is pruned to about 20 feet, it can provide an effective backdrop for lower-growing shrubs and perennials. The flowers are insignificant, but the glossy leaves turn bright yellow in the fall. In order to enjoy the unusual, large, and aromatic fruit, which can be messy when they fall off, it is necessary to have at least two plants, because male and female flowers grow on separate plants. Squirrels and black-tailed deer will eat the fruit, although it is not a significant source of food for wildlife. Osage oranges are also sold as roach and moth repellents.

CALLIOPSIS
Coreopsis tinctoria

Calliopsis

COREOPSIS TINCTORIA

IN JUNE AND JULY OF 1804 the explorers worked their way up the Missouri River across present-day Kansas and Nebraska. They must have seen coreopsis mingling with other wildflowers of the Great Plains that put on a colorful display during the summer months. Most of the extant journal entries of that time were written by William Clark. For the most part, Clark expressed less appreciation of botany than his cocaptain did, but he commented a number of times on the beautiful prairie. He wrote of how nature exerted herself to beautify the scenery by the variety of flowers rising delicately above the grass, which strike and perfume the senses and amuse the mind. He went on to wonder, in a manner characteristic of the time, why so magnificent a scene was created so far removed from civilization to be enjoyed only by buffalo, elk, deer, bear, and Indians.

American Indians, who undoubtedly enjoyed the beauty of the prairie flowers, also found uses for many of them. Coreopsis, for example, was used to make a dye. The Sioux, as the Lakota and their kin were known to the Corps, made a tea from coreopsis, but the expedition would not likely have learned of this, since their encounters with the Sioux were mostly rather tense and unfriendly.

Although Thomas Jefferson enjoyed flowers and gardening, he did not specifically instruct Meriwether Lewis to collect plants of ornamental value. There is a specimen in the Lewis and Clark Herbarium thought to be coreopsis but it is too poor to conclusively identify. The journals contain no description of the plant.

It was the British-born botanist, ornithologist, and printer Thomas Nuttall, who spent thirty-some years in North America, who described *Coreopsis tinctoria* for science and introduced it to gardeners. Soon after he arrived in the United States in 1808, he set out for the West Coast with a fur-trading group known as the Astorians, following the route of Lewis and Clark. Nuttall did not reach the Pacific on this trip, but he did gather a number of plants and was eager to go west again in 1819. This time journeying into the Arkansas territory, he again collected many plants, including *Coreopsis tinctoria*. Like others before and after, he was amazed by the vast and beautiful blooming open spaces, which he described in his *Journal of Travels into the Arkansas Territory*.

> *These vast plains, beautiful almost as the fancied Elysium, were now enamelled with innumerable flowers, among the most splendid of which were the azure Larkspur, gilded Coreopsides, Rudbeckias, fragrant Phloxes, and the purple Psilotria. Serene and charming as the blissful regions of fancy, nothing here appeared to exist but what contributes to harmony.*

Nuttall brought back from this trip seeds of various plants, including *Coreopsis tinctoria*, which he passed on to nurserymen and customers in the United States and England. It became known as Nuttall's weed or calliopsis; today it is often called tickseed or plains coreopsis. A dark-flowered variety had been introduced by mid-nineteenth century; according to one catalog, it deserved a place in every flower garden.

This species of coreopsis is still a popular garden flower, although it is regarded unfavorably by some in agribusiness. Because of its ability to easily self-sow, it has earned a reputation as a troublesome weed and therefore an enemy to be eradicated. Those who work with natural fibers and use coreopsis as a dye take a different view.

Calliopsis is a highly adaptable native wildflower that successfully made the transition to popular and colorful garden ornamental. Its bright yellow flower with mahogany center appears from late spring to early fall. Its natural range is extensive, including temperate regions of the continent west of the Mississippi River. Calliopsis blankets the prairies and grasslands in early summer and may still be putting out blooms in October in more northern parts of its range.

Calliopsis is easy to grow from seed and has a long bloom period, especially if deadheaded or sheared. Dwarf varieties are less prone to flopping. It is usually an annual but may return for a second year in certain conditions. The flowers are about an inch across and come in various combinations of yellow and mahogany; height ranges from less than a foot to four feet. Calliopsis prefers hot, dry, sunny locations.

BUR OAK
Quercus macrocarpa

Bur Oak

QUERCUS MACROCARPA

IN SEPTEMBER 1804 Lewis wrote about a species of white oak that evidently was not familiar to him in spite of his extensive experience in the American forest east of the Mississippi. Bur oak is more common in the central and northern regions of North America. When Lewis described it, the expedition was in South Dakota.

On arriving at their camping place on September 16, Lewis wrote that in addition to cottonwood, elm, and ash was a small species of white oak

> *which is loaded with acorns of an excellent flavor very little of the bitter roughness of the nuts of most species of oak, the leaf of this oak is small pale green and deeply indented, it seldom rises higher than thirty feet is much branched, the bark is rough and thick and of a light colour; the cup which contains the acorn is fringed on it's edges and imbraces the nut about one half; the acorns were now falling, and we concluded that the number of deer which we saw here had been induced thither by the acorns of which they are remarkably fond. Almost every species of wild game is fond of the acorn, the Buffaloe Elk, deer, bear, turkies, ducks, pigegians and even the wolves feed on them.*

Lewis's observations were on the mark, as they would be throughout the expedition. He identified several distinguishing characteristics of bur oak, a species that can be quite variable in some aspects. The shaggy cup may cover up to three-quarters of the nut and accounts for several of its common names, such as mossycup oak and overcup oak. The deeply indented leaves are also typical of bur oak. The rough, light-color bark is a feature of the white oak group of species. Acorns of most oaks are too high in tannins for human consumption unless they are leached, but Lewis apparently was curious and discovered that the bur oak acorn is sweet and edible. Years later, in 1898 when the botanist and horticulturist Thomas Meehan studied the Lewis and Clark collection for the Academy of Natural Sciences (established in 1812, in Philadelphia), he wrote that "the travelers were not botanists, but a specialist could not have drawn a better description of *Quercus macrocarpa*."

In the academy's Lewis and Clark Herbarium are two sheets of pressed and dried bur oak specimens. They represent several stages of growth, indicating that they were collected at different times of the year. Lewis collected bur oak in late summer, after the flowering season. There is also a specimen with catkins, the yellowish-green flowers that appear in the spring. Experts at the academy suggest that the catkin stem may have been collected in the spring of a later year by Thomas Nuttall and placed on the same sheet with Lewis's material.

The expedition probably passed bur oaks in a number of spots as they proceeded on their arduous way up the Missouri River. The species is found in a wide range of environments and, in moist, rich soils, can grow to great heights; in the drier, harsher conditions of the prairie, where Lewis found it, it may only reach the size of a shrub or small tree.

Bur oak is drought and fire resistant, two important qualifications for an inhabitant of the prairie. The thick bark that Lewis observed helps bur oak survive fire, a common natural occurrence in grasslands. In today's world, where fire suppression is the norm, bur oaks may be less successful. Without the regular occurrence of fire, less fire-resistant species grow alongside the bur oak, creating a shady situation in which bur oak saplings cannot survive.

Like many oak species, bur oak produces a heavy crop of acorns every few years. Judging by the great number of animals that Lewis observed eating the acorns, 1804 was a year of heavy production. The alternation of light and heavy seed production is thought to be a survival mechanism for the tree. In years of light production all the acorns are probably eaten by animals, but in heavy years a few escape and live to produce a new generation of trees.

The French botanist André Michaux was the first to describe bur oak for science. He traveled extensively in North America for eleven years at the end of the eighteenth century, collecting and writing about plants from Canada to Florida east of the Mississippi River. He saw bur oaks that reached heights of 60 to 70 feet. He admired the form of the tree, and having grown up on the edge of Versailles, he thought it would be an attractive addition to gardens and parks. He sent acorns of various oaks home to France, where they were germinated and studied with much interest. After returning to France in 1796, Michaux wrote a major work on the oaks of North America, which was beautifully illustrated by Pierre-Joseph Redouté, the famed botanical artist.

André Michaux and Meriwether Lewis are also linked through Thomas Jefferson. While he was minister to France, Jefferson had approached Michaux in January 1793 on behalf of the American Philosophical Society about leading a transcontinental expedition to the Pacific Ocean. Political intrigues intervened—certain French interests were organizing forces to seize the vast Louisiana territory from Spain—and when they were exposed, support for Michaux's expedition dried up. Eleven years later Lewis leapt at the opportunity.

Bur oak may be used in urban settings because it is resistant to pollution and does well in dry conditions. It is a slow-growing tree and may live for more than 300 years, as long as it is not in a wet site. It may also be planted along the edges of exposed areas to protect against severe wind and provide shelter for animals. The hard, durable wood is used in construction and furniture making.

NARROW-LEAF CONEFLOWER
Echinacea angustifolia

Narrow-Leaf Coneflower

ECHINACEA ANGUSTIFOLIA

THE LEWIS AND CLARK EXPEDITION spent the winter of 1804–05 at Fort Mandan, in North Dakota. In the spring the captains sent a large shipment down the Missouri to St. Louis, addressed to President Thomas Jefferson in Washington. They enclosed journals, field notes, and maps; astronomical observations, ethnographic information, a "statement of rivers, creeks, and most remarkable places"; boxes containing animals' skins and bones; samples of minerals; and cages with four live magpies and a prairie dog. Box four contained over sixty plant specimens, including bur oak and coneflower.

Many of the specimens that Lewis shipped to St. Louis that spring were labeled as to where and when they had been collected. This is not the case with the coneflower, so there is no way of knowing where or even if the explorers ever noted it in bloom or if indeed they gathered seeds of it. There is no specimen in the Lewis and Clark Herbarium.

The winter had been extremely harsh, with heavy snows and temperatures reaching forty degrees below zero. During one of the darkest days, William Clark wrote a journal entry regarding a root used by the Indians for curing the bite of a mad dog. Later entries expanded on the use of the plant to include curing rattlesnake bite. By March, when most of the snow had melted and the expedition was assembling the shipment to St. Louis before heading west, Clark collected a root of this plant, which was the narrow-leaf purple coneflower.

The captains learned about coneflower not from the Indians but from a Mr. Hugh Heney of the North West Company, who was trading with the Sioux. Mr. Heney met Lewis and Clark that winter among the Mandan Indians. In February 1805 Clark wrote that Mr. Heney brought letters, bearberry, and the root and top of a plant for the cure of snake- and dog bites. He gave a brief description of how to use it and wrote that it was found on high lands and sides of hills.

In a letter to Thomas Jefferson written in March, Lewis gave a detailed description of how to use the plant, a specimen of which was also enclosed in the shipment home.

> *This specimen of a plant common to the praries in this quarter was presented to me by Mr Hugh heney, a gentleman of rispectability and information who has resided many years among the natives of this country from whom he obtained the knowledge of it's virtues. Mr. Heney informed me that he had used the root of this plant frequently with the most happy effect in cases of the bite of the mad wolf or dog and also for the bite of the rattle snake. He assured me that he had made a great number of experiments on various subjects of men horses and dogs particularly in the case of madness, where the symptoms were in some instances far advanced and had never witnessed it's failing to produce the desired effect.*

Lewis went on to explain how to make a poultice from the root for external application, how the bitten place might have to be lacerated if it had healed over, and that best results would be obtained when applied early. This unfamiliar plant might have tremendous value, so he told Jefferson he was enclosing a few pounds of the root for the American Philosophical Society to experiment with.

Jefferson was evidently interested in a plant that could cure rattlesnake bites. Months after receiving the shipment, he consigned a packet of seeds to a prominent Philadelphia gardener, saying that they came from a plant "used by the Indians with extraordinary success for curing the bite of the rattle snake & other venomous animals."

The expedition apparently never had a need for echinacea, because at no time did they record an encounter with poisonous snakes or mad dogs. A member was bitten by a snake on July 4, 1804, but it did not turn out to be poisonous. About that time one of the expedition's interpreters told Clark of an extraordinary snake that gobbled like a turkey and could be heard several miles away; Clark did not seem to take the report seriously.

Echinacea angustifolia is a wildflower native to the North American prairie. It was much valued by Indians of the Great Plains, including the Blackfeet, Dakota, Pawnee, and Teton Sioux, all tribes that the expedition encountered. They used the root to treat not only venomous bites but also toothache and numerous other ailments. Today it is accepted by many as effective in treating cold and flu symptoms, infections, and slow-healing wounds and is used medicinally throughout much of the world. One of the most widely used herbal remedies in the United States and Europe, echinacea is also accepted by practitioners of Ayurvedic medicine, a form of healing practiced in India since ancient times.

The popular garden ornamental *Echinacea purpurea,* purple coneflower, is native to moister regions of the Midwest and Southeast. Like its narrow-leaved western relative, it also was used by Indians to treat health problems. The two species are similar, but the narrow-leaf is better adapted to severe drought conditions where rain may not occur for many months. The hairy stems, narrow leaves, and long thick taproot are characteristics that enable it to survive for long periods without water.

Narrow-leaf coneflower is shorter than purple coneflower. Its smaller flower is pale pink to pale lavender, making it a more delicate, less showy presence. It blooms in June and July and is a favorite of butterflies. It doesn't transplant well and may take several years to become established.

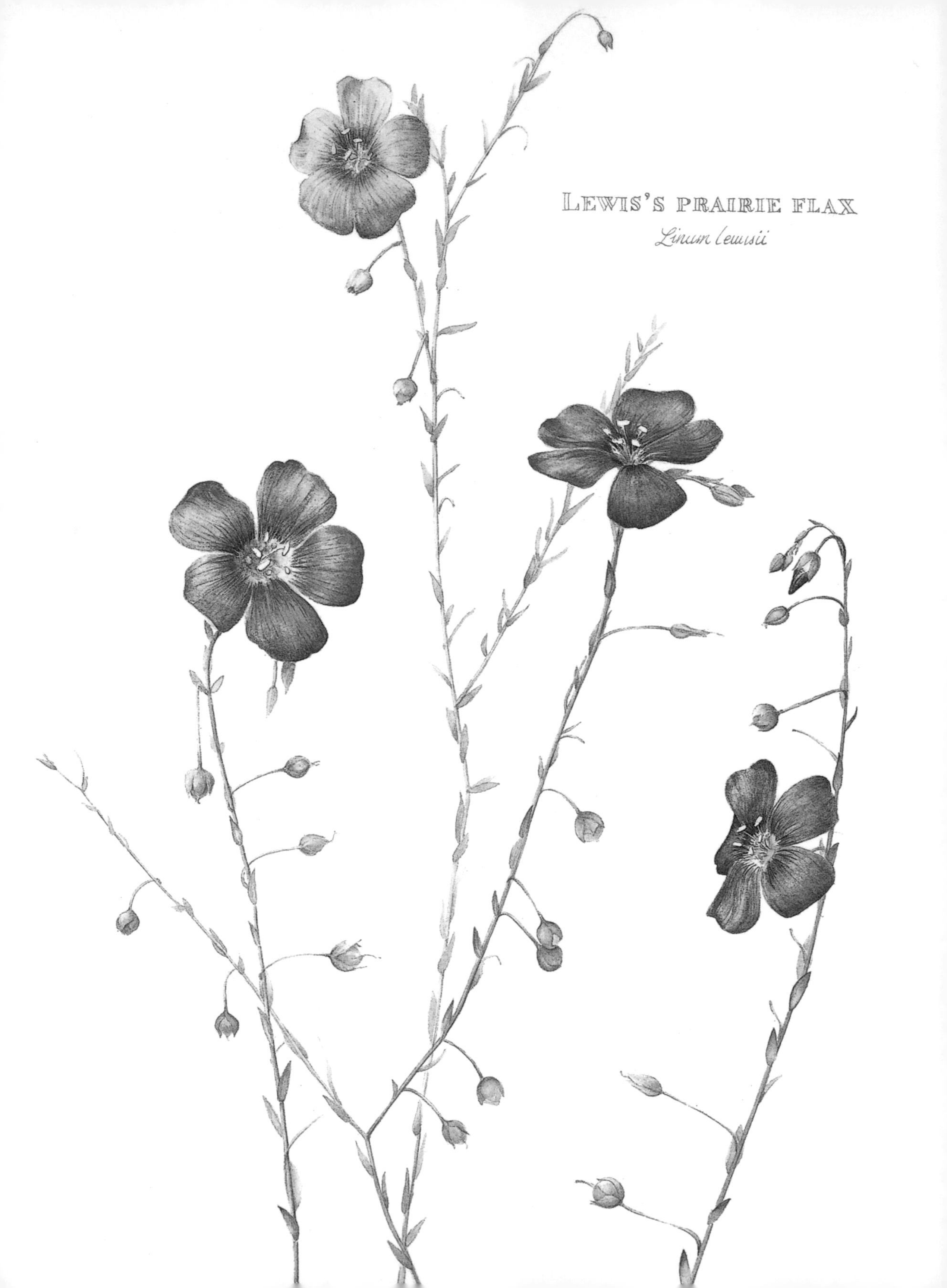
LEWIS'S PRAIRIE FLAX
Linum lewisii

Lewis's Prairie Flax

LINUM LEWISII

WHEN THE EXPEDITION STARTED working its way up the Missouri River in April 1805, the explorers were expecting to complete most of the trip by the following winter. On the last day of March Lewis sent his mother a letter in the shipment downriver to St. Louis. He wrote with optimism that he expected to reach the Pacific Ocean that summer and be halfway home by winter. As the months wore on, the men began to see mountains in the far distance, which filled them with awe increasingly tinged with apprehension about what lay ahead. Nevertheless, during the trek across the plains and foothills in the spring and summer months, they marveled at the beauty of the landscape, the spectacular animals, including huge herds of buffalo, gentle and approachable antelope, and agile mountain goats only to be glimpsed in the distance, and numerous plants both familiar and unknown.

In April Lewis described hills covered with aromatic herbs that resembled in taste, smell, and appearance sage, hyssop, wormwood, southernwood, and with two plants that were new to him. One resembled camphor. The other had a long, narrow, smooth, soft leaf with an agreeable smell and flavor. He wrote "of this last the Atelope is very fond; they feed on it, and perfume the hair of their foreheads and necks with it by rubing against it."

On June 12, Lewis recorded seeing several beautiful and picturesque ranges of mountains, the most distant of which were snow covered, "an august spectacle and still rendered more formidable by the recollection that we had them to pass." Around that same time, two other members of the expedition, Sergeant John Ordway and Sergeant Patrick Gass, recorded seeing great quantities of a blue-flowered plant that resembled flax. Although it is not certain that Ordway and Gass saw flax, on July 18 Lewis without doubt described a species of flax not previously known to Western science.

> *I have observed for several days a species of flax growing in the river bottoms the leaf stem and pericarp of which resembles the common flax cultivated in the U' States. the stem rises to the hight of about 2½ or 3 feet high; as many as 8 or ten of which proceede from the same root. the root appears to be perennial.*

Flax was, of course, a highly useful plant both for the strong fibers of its stems, which were made into linen, and for the oil of its seed. The Old World species of flax, used in Europe and the Middle East since antiquity, is an annual plant. It is still the one most widely used.

Lewis further described his flax as having thick, strong bark and growing from a single root. He thought this flax might be perennial and hoped to collect the seed at a later date, which might "on experiment prove to yeald good flax and at the same time admit of being cut without injuring the perennial root." This would make it a valuable plant

> *for notwithstanding the seed have not yet arrived at maturity it is puting up suckers or young shoots from the same root and would seem therefore that those which are fully grown and which are in the proper state of vegitation to produce the best fax are not longer essencial to the preservation or support of the root.*

In other words, a flax grower would not have to replant each year. He could harvest mature stems and seeds and get new growth from the root for a future crop. Although various Indian tribes used the native flax for food, fiber, fishing nets, and cords, there is no indication in the journals that the explorers ever learned of these uses.

Several days later, in a spot where "high plains succeeds the river bottoms and extend back on either side to the base of the mountains," Lewis found ripe flax seeds and collected them. As the expedition advanced into the high mountains and approached the Continental Divide, of necessity they changed their mode of transport from boat to horse. They made contact in August with the people of Sacagawea, the Shoshone woman who accompanied the Corps. The Shoshone possessed large herds of horses and were willing to trade and to offer advice on how to cross the mountains. To lighten the expedition's load, Lewis ordered the construction of a cache in which to store items to be collected on the return trip. Among the items he placed in the cache were specimens of plants, minerals, and mostly likely the flax seed. Returning the following summer, he found that all the plant specimens in the cache had been destroyed by rising water from the nearby Missouri River. Fortunately, he collected more seeds and a specimen of the plant this time.

A year after the return of the expedition, Bernard McMahon reported to Thomas Jefferson that he was having success growing the perennial flax and several other plants collected by Lewis. Although Lewis's prairie flax did not prove to be useful for either fiber or oil, it became a popular garden ornamental later in the nineteenth century.

Linum lewisii was given its scientific name a few years after Lewis's death in 1809. The binomial system of plant nomenclature, designed by Carolus Linnaeus in the eighteenth century and still in use today, was becoming established in the early nineteenth century. Lewis, who was the first to describe the plant, was not a trained botanist nor, probably, was he aware of the correct procedures for being recognized as the "author" of a newly discovered plant. However, in this instance the plant was named by a young German botanist, Frederick Pursh, who had the decency to recognize Meriwether Lewis when he named this potentially important plant. Pursh had the opportunity, beginning in 1807, to review much of the Lewis and Clark plant collection, and his name has been inextricably woven into the story ever since. His major work on the flora of North America was extensively based on the Lewis and Clark specimens.

Lewis's prairie flax is a pretty garden ornamental suited to hot, dry sites. Each morning delicate sky blue flowers open on slender arching stems, only to fall off in the afternoon and be replaced by others the next morning. In spite of its fragile appearance, it is quite sturdy and may put out a second flush of blossoms on new growth in late summer.

LEWIS WROTE A DETAILED DESCRIPTION OF THIS NEW SPECIES OF PERENNIAL FLAX, *LINUM LEWISII*, IN THE SUMMER OF 1805 WHILE HEADING UP THE MISSOURI RIVER. TODAY IT IS A POPULAR GARDEN ORNAMENTAL. EACH SKY BLUE FLOWER LASTS JUST ONE DAY, BUT A NEW ONE FARTHER OUT ON THE STEM OPENS THE NEXT DAY.

PLAINS PRICKLY PEAR
Opuntia polyacantha

Prickly Pear

OPUNTIA SPP.

PRICKLY PEARS are mentioned numerous times in entries by both Captain Lewis and Captain Clark, and for good reason. As Lewis wrote at one point, referring to mosquitoes, gnats, and prickly pears, "our trio of pests still invade and obstruct us on all occasions." In a rare reference to the Bible, he went on to say that they were equal to any three curses that Egypt ever labored under.

During the summers of 1804 and 1805, all members of the party suffered mightily from stepping on prickly pears. At one time Clark extracted seventeen spines from his feet. The soles of their moccasins were not thick enough to keep the spines out, even though they added extra layers of leather. Lewis wrote that progress was slowed because the pears were so abundant the men had to give half their attention to where they stepped. At times they found little space free of pears where they could lie down at night. Compounding their misery were the sharp stones that cut their feet and a bearded grass that also caused Lewis's dog great pain. Fortunately, on the return trip in the summer of 1806, they rode their boats downriver with the current and did not often travel by land.

In mid-June 1805 Lewis and a small advance party were exploring ahead of the rest of the expedition. They were looking for the Great Falls of the Missouri that would prove that they had not followed a tributary by mistake. While the rest of the party were taking care of matters at camp, Lewis went off on his own. Intending to be back for dinner, he found himself irresistibly drawn farther ahead by the roar of cascading water and the increasingly spectacular falls. He noted a beautiful little island in the middle of the river, where an eagle had built a nest in a cottonwood tree. On arriving at a height overlooking a "most beatifull and extensive plain reaching . . . to the base of the Snowclad mountains," he decided to kill a buffalo for food in case he could not get back to camp that night. While watching the animal die, he forgot to reload his rifle and found himself being charged by an immense grizzly bear. He had no choice but to run into the river. When he faced the animal in an "attitude of defence" the bear "declined the combat" and ran full speed in the opposite direction as if pursued until it disappeared from Lewis's sight. The bear's mysterious flight gave Lewis confidence to continue. When he finally looked at his watch, it was 6:30 in the evening. He decided to turn back, only to run into an animal that he first thought to be a wolf. But it was brownish yellow, crouched like a cat, and when Lewis fired his gun, it disappeared into a burrow, though he was "almost confident" that he had struck the creature. Lewis next encountered three buffalo that separated from the herd and galloped toward him only to stop and turn away.

> *I then continued my rout homewards passed the buffaloe which I had killed, but did not think it prudent to remain all night at this place which really from the succession of curious adventures wore the impression on my mind of inchantment; at sometimes for a moment I thought it might be a dream, but the prickley pears which pierced my feet very severely once in awhile, particularly after it grew dark, convinced me that I was really awake.*

Prickly pear, a member of the cactus family, is native to those parts of North America where the climate is dry and the soil sandy. It seems to have been familiar to the explorers, who probably had encountered other species of *Opuntia* prior to their westward trip. In spite of the extreme nuisance they represented, Lewis was able to enjoy them. In July 1805 he wrote that the prickly pear was now in full bloom and formed one of the beauties of the plains. Clark's corresponding entry made no comment about beauty.

On August 13, 1805, just five days before his thirty-second birthday, Lewis had his first meeting with the Shoshone. In addition to recording this significant event, he observed that they had now encountered three species of prickly pear. One was broad-leaved and common to the Missouri; another was globular and common to the upper Missouri, especially once it entered the Rocky Mountains. The third was "peculiar to this country," that is, the Bitterroot Mountains.

> *it consists of small circular thick leaves with a much greater number of thorns. these thorns are stronger and appear to be barbed. the leaves grow from the margins of each other as in the broad leafed pear of the missouri, but are so slightly attatched that when the thorn touches your mockerson it adhears and brings with it the leaf covered in every direction with many others. this is much the most troublesome plant of the three.*

The three species that Lewis noted are thought to be *Opuntia fragilis*, *O. macrorhiza*, and *O. polyacantha*. In a bow to botanical correctness, it must be noted that *spine* is the proper term, not *thorn* or *briar*, as the captains wrote.

In mid-October Captain Clark observed an Indian lodge where fish and prickly pears were drying, the pears to be used as fuel come winter, and several times the party saw animals eating the pears. Not surprisingly, they did not collect a specimen of prickly pear to bring home, nor were they ever tempted to sample them as comestibles.

WESTERN SERVICEBERRY

Amelanchier alnifolia

Western Serviceberry

AMELANCHIER ALNIFOLIA

THE EXPEDITION DELIGHTED in eating various small fruits when available, which they compared to familiar ones from home. In addition to serviceberries, Lewis noted gooseberries, chokecherries, cranberries, nannyberries, huckleberries, strawberries, haws, and purple, red, black, and yellow currants. Serviceberries are found across northern North America and were observed by various members of the expedition along much of the way. By the time they reached the mountains in the summer of 1805, they were seeing a species new to Western science known today as *Amelanchier alnifolia,* western or Saskatoon serviceberry.

In late July the expedition came to a fork formed by three rivers. The captains knew they had reached the end of the Missouri and named the three rivers after Thomas Jefferson, James Madison, and Albert Gallatin, the president and the secretaries of state and treasury. Timber and game were becoming scarce; the soil, however, was quite rich and grass was abundant.

On August 2, while working their way up the Jefferson River, Lewis wrote of suffocating, intense heat at midday and nights so cold that two blankets were insufficient covering. He also described feasting on various fruits.

> *we found a great courants, two kinds of which were red, others yellow deep purple and black, also black goosburies and service buries now ripe and in full perfection, we feasted suptuously on our wild fruit particularly the yellow*

courant and the deep purple servicebury which I found to be excellent the courrant grows very much like the red currant common to the gardens in the atlantic states tho' the leaf is somewhat different and the growth taller. the service burry grows on a smaller bush and differs from ours only in colour and the superior excellence of it's flavor and size, it is of a deep purple.

Serviceberries had been eaten by settlers since colonial times. Native Americans dried them for winter food and added them to cakes and to pemmican made from meat and fat. They used the wood for arrows, the fibers for cordage, and the leaves and twigs for medicines.

The fruit of eastern species, at least by some accounts, is rather flavorless. Lewis might not have commented as he did on August 2 if western serviceberries had not been so noticeably tasty. The flavorful western berry, native from the Dakotas to the Pacific coast, was first described for science by Thomas Nuttall a decade after Lewis encountered it.

On August 12 Lewis described their sense of accomplishment at being able to stand astride the smallest remaining tributary of the "mighty and heretofore deemed endless" Missouri River. They crossed the Continental Divide and for the first time drank the westward-flowing water of the Columbia River watershed. At the top of the dividing ridge Lewis had "discovered immence ranges of high mountains still to the West of us with their tops partially covered with snow." Although the sight of these mountains must have been a shock, since they were unanticipated, the expedition had more immediate matters to attend to. At this point they had not yet met up with Shoshone Indians from whom they could obtain horses for the overland trek.

The next day Lewis did make contact with the Shoshone, and perhaps sensing the momentousness of the occasion, he wrote a lengthy journal entry. He mentioned a number of plants, including the troublesome prickly pear and the snowberry. The Indians offered them tobacco and cakes made of dried serviceberries and chokecherries, which was all they had to eat. From this time on the expedition would survive on a diet mainly of berries, roots, and fish, rather than game, which they much preferred. Lewis soon had his first taste of salmon, which further convinced him that he was on waters that flowed to the Pacific.

The following spring, after a soggy winter on the Pacific coast, the expedition set out for home and this time saw the serviceberry in bloom. Several specimens in the Lewis and Clark Herbarium are thought to be *Amelanchier alnifolia*. One includes a flower and was collected on April 15, 1806, on the narrows of the Columbia River; the others were collected in May and June farther along on the homeward route.

In the East a common name for this shrub is shadblow, so named because it blossoms at the time the shad return from the ocean to the rivers to spawn. Whether or not the expedition were familiar with this East Coast phenomenon, there was no similar occurrence on the western coast. They noted the serviceberry blooming in early April and were told by the Indians that the salmon would not arrive on the river before early May.

The fruit of the western species starts out red and becomes dark purple when ripe. Serviceberry is a member of the rose family, and the fruit is not technically a berry but a pome, like a miniature apple. It is a shrub or small tree, and its early white blooms are a welcome harbinger of spring. Wildlife eat the fruit and browse on the leaves and twigs.

SNOWBERRY
Symphoricarpos albus

Snowberry

SYMPHORICARPOS ALBUS

SNOWBERRY BELONGS to the honeysuckle family and is native to the western mountains. Lewis referred to it a number of times simply as the white-berried honeysuckle. During the course of the expedition he noted other members of this family as well, including wolfberry, which is commonly found on the Plains and which he collected while traveling up the Missouri River.

It is not known if he carried stems or leaves of snowberry back to the United States, but he almost certainly did take some seeds. In mid-August 1805, when the expedition was getting into the range of snowberry, Lewis wrote about the plant and its distinctive berries for the first time.

> *a species of honeysuckle much in it's growth and leaf like the small honeysuckle of the Missouri only reather larger and bears a globular berry as large as a garden pea and as white as wax. This berry is formed of a thin smooth pellicle which envellopes a soft white musilagenous substance in which there are several small brown seed irregularly scattered or intermixed without any sell or perceptable membranous covering.*

By mid-September the Corps was toiling through high mountains. Food was scarce and this was a difficult time for the expedition. Clark, who was proceeding ahead of the main party with a group of hunters, came to a watercourse that he named Hungry Creek. The terrain was covered with great quantities of fallen timber, and the pack horses kept slipping and falling on the steep slopes. There was little grass for the horses, so they were allowed to roam at night, and as a result the explorers had to spend time searching for them in the mornings. On September 20 Lewis recorded that the horse carrying winter clothing had gone missing. In spite of these tribulations, he observed a number of new birds and plants and commented again on the snowberry, "a kind of honeysuckle which . . . rises about 4 feet high not common but to the western side of the rocky mountains."

In 1812, six years after the return of the expedition and two and a half years after Lewis's death, snowberry was mentioned in correspondence between Thomas Jefferson and Bernard McMahon. McMahon was sending plants to Jefferson, including *Symphoricarpos,* "a beautiful shrub brought by C[aptain] Lewis from the River Columbia." He explained that its berries hang in large clusters and are snow white; he had given it the common name of snowberry bush. The following year, Jefferson sent cuttings of snowberry to his friend Noailles de Tessé, in France. He also grew snowberry at Monticello and in 1826, shortly after he died, his granddaughter Cornelia Randolph wrote to her sister about some of the plants still growing there, including her "favorite snowberry, so light and elegant in its form and foliage, and its berries so beautiful and pure, but most valued to me because it most flourishes when all other flowers have faded."

No pressed specimen exists in the Lewis and Clark Herbarium, but there is one in the Charleston Museum in South Carolina. Only in the past decade was a connection made between this specimen and the Lewis and Clark expedition. The Charleston sheet is associated with two enthusiastic plant collectors who were well acquainted with the Philadelphia group of naturalists who were avidly interested in the Lewis and Clark plant collection. The question for botanists to resolve is whether the Charleston specimen is indeed one collected by Lewis or a cutting of a plant grown in Philadelphia from seed collected by Lewis.

The scientific name *Symphoricarpos* refers to the tightly clustered fruit. Snowberry is a dainty deciduous shrub, three to six feet tall, with slender branches that arch over to the ground. The smooth, reddish branchlets seem too delicate to support the heavy weight of the ripe clusters of berries that appear in late summer and last into fall. When the plant is flourishing, the pure white berries can be so profuse that they resemble drifts of snow. They seem to have little nutritional value or flavor that would appeal to either wildlife or humans. The dark green leaves are about an inch long and contrast nicely with the white berries. McMahon, Jefferson, and Cornelia Randolph all noted how beautiful the berries looked in winter.

The flowers are small and inconspicuous. Snowberry tolerates sun and partial shade and can be propagated from seeds, cuttings, and suckers. The soil of its native habitat in the western mountains, where it thrives, contains limestone and clay. By the middle of the nineteenth century it was commonly found in gardens in the eastern part of the country; it is available today in garden centers. Snowberry is sometimes planted on banks to hold soil in place and prevent erosion.

ANGELICA
Angelica arguta

Angelica

ANGELICA SP.

"ALL GONE!" These words on a label attached to the Lewis and Clark Herbarium sheet of paper tell the unfortunate fate of the angelica specimens. Written probably in the late nineteenth century, they succinctly state that nothing remains. Whether eaten by insects or destroyed some other way, in the years following the return of the expedition to Philadelphia, these specimens had become dust.

Two other labels are still on the sheet that once contained the angelica. These were written by Frederick Pursh, the German botanist in Philadelphia who was asked to describe the plants soon after the expedition's return in 1806. Pursh most likely was rewriting notes written by Lewis. They read "Angelica within the Rocky mountains in moist places, June 25th 1806" and "The flowering one taken in September 3rd 1805."

In his journals Lewis said very little about the plant. He made no comment on or around September 3, 1805, regarding the "flowering one," although interestingly enough, on September 4, Sergeant Patrick Gass wrote the following in his journal:

> *We ... proceeded down a small valley about a mile wide, with rich black soil; in which there are a great quantity of sweet roots and herbs, such as sweet myrrh, angelica, and several others, that the natives make use of, and of the names of which I am unacquainted.*

Nor can we be sure of the identity of the plants that Gass saw. Sweet myrrh might be sweet cicely, a familiar plant brought from Europe to

American gardens; it has a western relative native to the Rocky Mountains. Much the same can be said for angelica. He was noting plants that, if not the same species, were closely related to ones familiar to him from home.

Gass's comment of September 4 suggests a more pleasant circumstance than was actually the case. The weather and the traveling conditions at that moment were extremely harsh. The air was cold and thin because of the altitude, snow had fallen during the night, and the trail was steep and slippery. The captains were deeply concerned about making it out of the mountains before winter, as food was becoming scarce and harsher weather lay ahead. It was apparent by then that they would not make it to the Pacific Ocean and back across the mountains that year.

On the return trip the following June, when the expedition was once again in the Bitterroots, Lewis observed

> *there is an abundance of a speceies of anjelico in these mountains, much stonger to the taist and more highly scented than speceis common to the U' States. know of no particular virtue or property it possesses; the natives dry it cut it in small peices which they string on a small cord and place about their necks; it smells very pleasantly.*

Later that month, at least according to the Herbarium label, he collected a specimen of angelica in a moist place. The expedition was near Hungry Creek, where Lewis had collected snowberry the preceding fall. He did not describe the angelica, although he recognized it as an unfamiliar species of a familiar plant. It is impossible to say which of the several species native to the Northwest it might have been.

Angelica was valued by both Europeans and American Indians for a number of medicinal uses, including respiratory ailments, and as a sweetener. During the course of the expedition, Lewis and other members made observations of additional plants that we think of as herbs—that is, plants that are useful medicinally, for flavoring, for repelling insects, or for their pleasant scents—including tansy, horsemint, achillea, artemisia, hyssop, and sage.

Along with his other duties, Captain Lewis was the one most often called upon to tend the sick and injured. The expedition was well stocked with medical supplies, including Peruvian bark (cinchona), which contains quinine for malarial fevers; laudanum (tincture of opium) for killing pain; mercury, one of the earliest chemical treatments for syphilis; and fifty dozen tablets of Benjamin Rush's "Thunderbolt," a popular laxative of the day. They carried syringes, lancets for bloodletting, lint for dressings, and other medical supplies and equipment. Before the start of the expedition Lewis had visited Dr. Rush in Philadelphia for an intensive course in medical care. However, Lewis had another weapon in his arsenal for combating health problems: a knowledge of herbal remedies. His mother, Lucy Meriwether, was well known throughout Albemarle County, Virginia, for her able use of herbs and traditional simples. A simple, a single herb used as a complete cure, might or might not work, but would be no more likely to harm the patient than some of the more "modern" treatments.

Lewis had learned about herbal remedies from his mother, and he was not averse to trying them when all else failed or the medicine chest was not handy. On one occasion he found himself suffering from violent intestinal pain and fever. He made a "strong black decoction of an astringent bitter tast" from chokecherry twigs that completely restored him within hours to his usual state of well-being.

Lewis and Clark treated not only their own men but also Sacagawea, as well as the Indians who accompanied them over the mountains in the spring of 1806. Once when Sacagawea was ill, they bled her and then tried other remedies when it didn't work. After nine days she recovered, either because of or in spite of her treatments. Except for Charles Floyd, who died early on, probably from appendicitis, all members of the expedition survived and made it back safely. Patrick Gass, the longest-lived member of the expedition, died in 1870 at age 99.

CAMAS
Camassia quamash

Camas

CAMASSIA QUAMASH

THE CORPS OF DISCOVERY first encountered *Camassia quamash* in the fall of 1805 as they were descending the western slopes of the Bitterroot Mountains in present-day Idaho. The explorers were hungry and weary; game and other sources of food were scarce, the country was rugged, and the weather was getting worse. William Clark was again leading an advance party in hopes of finding deer or other game. On September 20 he came upon many Indian lodges. The chief and most of the warriors were away, but one man led him to the chief's lodge, where he was given food, including bread made from camas roots, which he ate heartily. The Indians were members of the Nez Perce tribe, and camas root, or quamash, was a much liked and important source of food for them.

On September 22 Clark had rejoined Captain Lewis and the rest of the party, and the next day he wrote that they purchased from the Indians roots, dried, in bread, and raw, as well as berries of red haws and fish. That evening Captain Lewis and two of the men were very sick. Over the next few days nearly all the men were "complaining of their bowels, a heaviness at the stomach and lax." Evidently the sudden change in diet to camas roots, dried fish, and berries was the cause of the trouble. Up until a short time before, their diet had consisted almost exclusively of meat obtained from animals of the Great Plains, including buffalo, antelope, and elk. Over the course of the winter spent at Fort Clatsop on the

Oregon coast, they adapted to the new diet and consumed much of the camas root, although they never really enjoyed it.

The following spring they encountered the camas when it was in bloom, and a beautiful sight it must have been. The expedition, in an ebullient mood, was heading home. They waited impatiently through May and into June to cross the Bitterroot Mountains because the snow, although melting fast, was still too deep. At Weippe Prairie, Lewis wrote on June 12, 1806, that the camas was in bloom and "at a short distance it resembles lakes of fine clear water, so complete is this deseption that on first sight I could have swoarn it was water."

Just the day before, he had written an extensive description of the camas plant using much botanical terminology; Thomas Jefferson had prepared his protégé well. Lewis began by saying that they found the camas in or adjacent to pine- or fir-timbered country and always in open ground and glades. The specimens that he had observed growing nearer the Pacific Ocean appeared in smaller quantities and were of inferior size compared to those on high flats and valleys within the Rocky Mountains.

> *it delights in a black rich moist soil, and even grows most luxuriantly where the land remains from 6 to nine inches under water untill the seed are nearly perfect which in this neighbourhood or on these flats is about the last of this month....this bulb is from the size of a nutmeg to that of a hens egg.*

Of the flower he said "the corolla consists of six long oval obtusly pointed skye blue or water coloured petals, each about 1 inch in length." He also described the tunicate bulb, the numerous radicles, the solitary peduncle, and the anther, which bursts and discharges its pollen just a few hours after the corolla unfolds. Soon after the seeds mature, the foliage dies down, the ground becomes dry, and the root increases in size, and by mid-July the Indians begin to collect it. He described how they prepared it for use over the fall and winter. Properly cooked and dried camas roots can keep for many years. Lewis concluded, "this root is palatable but disagrees with me in every shape I have ever used it."

Perhaps the explorers were able to get some relief for their stomach problems caused by eating various kinds of roots. Sacagawea introduced them to a plant that Lewis thought might be a species of fennel. They found its taste very agreeable and that it had an added benefit, which was to "dispell the wind which the roots called Cows and quawmash are apt to create particularly the latter." On June 23, 1806, Lewis collected a specimen of *Camassia quamash* that safely made it back East and is now in the Lewis and Clark Herbarium of the Academy of Natural Sciences.

This lovely plant is a member of the lily family, although it is sometimes referred to as wild hyacinth. It forms tall spikes one to three feet tall, along which starlike pale blue to deep violet flowers come out in the spring. It is a perennial and, as the description by Lewis indicates, likes moist meadows and edges of ponds. It is native to the Pacific Northwest and east to Wyoming.

BEARBERRY
Arctostaphylos uva-ursi

Bearberry

ARCTOSTAPHYLOS UVA-URSI

THE EXPLORERS ENCOUNTERED *Arctostaphylos uva-ursi* in a number of places along the route. The low-growing shrub is indigenous across northern North America, Europe, and Asia. Although bearberry was not new to science, Lewis and Clark noted its presence a number of times and provided interesting ethnobotanical information, recording how it was used by various American Indian groups.

The specimen in the Lewis and Clark Herbarium was probably collected at Fort Mandan in North Dakota in late 1804. At that time they mentioned that the Teton Sioux used scrapings and shavings of it to blend with other materials, such as red osier dogwood, for smoking. The explorers referred to bearberry as saccacommis.

The scientific name *Arctostaphylos uva-ursi* is both Greek and Latin and tells us something about the plant. In Greek, *arctos* means bear and *staphyle* is a bunch of grapes. Similarly *uva* and *ursi* are Latin for grape and bear. Some of its many common names are kinnikinnick, chipmunk's apple, and hog cranberry. It is a food source for many wild animals, including deer, rodents, birds, and, of course, bears.

The expedition had a number of frightening encounters with grizzly bears, animals that at the time were new to science. In April 1805 they saw huge bear tracks around the carcasses of buffalos and looked forward to meeting the animal that had made them. With a degree of bravado Lewis described it as a formidable and furious animal that "the Indians may well fear . . . but in the hands of skillfull riflemen they are by no means as formidable or dangerous as they have been represented." They soon saw the strength of the grizzly bear, finding that it took numerous gunshots to bring one down and that it was as capable of killing them as they were of killing it. During the expedition's final homeward stretch in the summer of 1806, Lewis wrote "these bear are a most tremendous animal; it seems that the hand of providence has been most wonderfully in our favor with rispect to them, or some of us would long since have fallen a sacrifice to their farosity."

The explorers learned that the Indians also ate bearberries and that mixing saccacommis with their tobacco, which was in short supply, made a pleasant smoke. Lewis incorrectly thought the name *saccacommis* came from the French and referred to sacks in which the trading companies carried the bearberry leaves. It comes from a Chippewa word. Neither Lewis nor Clark ever arrived at a uniform way of spelling it, Clark being the far more creative. In one month alone he wrote *Sackacomma, Sackacomey, Sackay Commis, Sackacomie, Sackacome,* and *Sackey Commy.*

As the end of the year 1805 approached, the expedition at last reached the Pacific Ocean. They spent a wet winter on the Oregon coast under overcast skies, but at least it was not cold. On December 7 they selected a spot somewhat inland to build a shelter, which they named Fort Clatsop after a local Indian tribe. The next day Clark set out with a small party to find a place to "make Salt . . . and See the probibillity of game." He spent the night of December 9 in what sounds like pleasant surroundings. A young chief invited him to his lodge, where Clark was treated with great politeness. The chief and his wife gave him a new mat to sit on and produced fish, licorice, and black roots, cranberries, and saccacommis berries in bowls made of horn. They also gave him a soup "of bread made of berries common to this Countrey" served in a neat wooden trencher with a cockle shell to eat it with. While a tremendous storm raged outside, the Indians played a game with beans passed from hand to hand and another with pieces like those used in backgammon.

Lewis wrote a lengthy description of bearberry the following January. He drew on his memory of having seen it thriving in drier, mountainous habitats and compared it to specimens in his current location.

> *The Sac a commis is the growth of high dry situations, and invariably in a piney country or on its borders. it is generally found in the open piney woodland as on the Western side of the Rocky mountains but in this neighbourhood we find it only in the praries or on their borders in the more open wood lands; a very rich soil is not absolutely necessary, as a meager one frequently produces abundantly. the natives on this side of the Rockey mountains who can procure this berry invariably use it; to me it is a very tasteless and insippid fruit.*

He wrote that it was evergreen and gave a detailed description of the leaf. As for its growth habit, it was much branched; procumbent, not creeping; and put out radicles that served to anchor it to the ground but not to provide nourishment. He commented on the flaky bark and that the berry ripened in September and was not affected by frost. It was a fine scarlet color outside and the inner part consisted of a dry mealy powder. The berries were gathered by the Indians and hung in bags in their lodges, "where they dry without further trouble, for in their most succulent state they appear to be almost as dry as flour."

Bearberry is a member of the heath family, which makes it a relative of rhododendron, azalea, blueberry, pieris, and other familiar garden plants. It is a low, trailing shrub that forms mats several feet broad and a few inches tall. The small white to pink flowers in spring are followed by red berries. The evergreen foliage turns reddish bronze in the fall. This native of temperate regions of the Northern Hemisphere can be purchased in garden centers.

OREGON GRAPE HOLLY
Mahonia aquifolium

Oregon Grape Holly

MAHONIA AQUIFOLIUM

BY JANUARY 1, 1806, the Corps of Discovery had completed construction of Fort Clatsop and set up their salt works at the coast. Lewis and Clark were ready to establish a regular work routine, including hunting and making moccasins and other clothing from animal hides. A policy was put in place regarding times when Indians would be allowed within the fort. Both men knew they would be there for some months, because the Indians had advised them that the Rocky Mountains would not be passable before June.

Although both Clark and Lewis often commented that nothing worthy of note had happened on a particular day, the men were kept very busy. By mid-February Clark had completed a map of the country from the mouth of the Missouri River to the Pacific coast. All previous maps had been filled with errors and guesswork regarding the upper Missouri River and the Rocky Mountains. Clark, who was an excellent mapmaker, provided much long-sought information, producing more than 100 maps during the two and a half years of the expedition.

Throughout the duration of the expedition, the captains copied each other's journal entries as insurance against loss. While they were at Fort Clatsop, in addition to writing extensively about the natural history and Indian culture of the Pacific Northwest, Lewis and Clark illustrated their work. The journals of this period include sketches of arrows, fish-hooks, knives, canoes, paddles, birds, fishes, and plants. They seem to have copied each other's drawings very closely, Clark having perhaps a slightly crisper style that gave a bit more attention to detail.

One of Clark's drawings is of a compound leaf of *Mahonia aquifolium*, showing three paired and one terminal leaflet, all armed with spinelike teeth. The hollylike evergreen leaves are shiny as if wet, hence the name *aquifolium*. The shrub, which is native to the Pacific Northwest, puts out purple berries in summer. Depending on what source you consult, *Mahonia* is either in the same genus as barberry (*Berberis*) or is its own closely related genus.

On February 12 Lewis wrote about two species of *Mahonia*, both new to science.

> *There are two species of ever green shrubs which I first met with at the grand rappids of the Columbia and which I have since found in this neighbourhood also; they grow in rich dry ground not far usually from some watercourse. the roots of both species are creeping and celindric. The stem of the first [M. aquifolium] is from a foot to 18 inches high and as large as a goosqull; it is simple unbranced and erect. its leaves are cauline, compound and spreading. the leaflets are jointed and oppositely pinnate, 3 pare & terminating in one, sessile, widest at the base and tapering to an accuminated point, an inch and a quarter the greatest width, and 3 inches & a 1/4 in length. each point of their crenate margins armed with a subulate thorn or spine and are from 13 to 17 in number. they are also veined, glossy, carinated and wrinkled; their points obliquely pointing towards the extremity of the common footstalk.*

Lewis took with him a number of reference books, including an illustrated work on botanical terminology of Linnaeus. Another was Benjamin Smith Barton's *Elements of Botany*, the first botany textbook written in the United States, which he had purchased in Philadelphia. He had also borrowed from Barton a history of Louisiana; in 1807, having carried it across the continent, he returned it to its owner with a note of thanks. He seems to have found time to consult these works while at Fort Clatsop, judging from the detailed descriptions written during this period of forced inactivity. The other species that Lewis described in the same entry is *Mahonia nervosa,* dull Oregon grape.

He did not comment on the striking yellow flower of either species, which appears in winter or early spring; the Herbarium specimen does, however, include a portion of a flower. The fruit was used as food and medicine by the Indians.

There is another, less conspicuous species, *Mahonia repens,* that the expedition would have encountered in the mountains but did not mention. American Indians used it in the treatment of venereal disease, a subject that concerned Lewis. During the winter at Fort Clatsop he wrote the following in regard to Private Goodrich.

> *Goodrich has recovered from the Louis veneri which he contracted from an amorous contact with a Chinnook damsel....I cannot learn that the Indians have any simples which are sovereign specifics in the cure of this disease.*

The winter of 1806 was the beginning of a rich time for collecting and describing plants, many of them new to science. In just a couple of weeks' time Lewis wrote about an edible thistle, three species of fern, a horsetail, salal, false solomon's seal, Oregon crabapple, vine maple, Pacific blackberry, evergreen huckleberry, and Oregon grape holly. In the coming months many more new species would be recorded.

About a decade later, Thomas Nuttall, the British botanist who spent most of his botanizing years in North America, named the Oregon grape holly in honor of Bernard McMahon, who had grown it from seeds brought back east by Lewis. By mid-nineteenth century it had become a highly prized nursery plant sold in catalogs.

In the upper right of the Herbarium sheet containing the Oregon grape holly specimens is a stamped annotation: EX HERB. A. B. LAMBERT. Frederick Pursh was residing with Bernard McMahon in 1809 and preparing formal descriptions of Lewis's plants. Lewis never returned to

Philadelphia to work with Pursh. A short time later Pursh left the country, taking with him a number of the plants that Lewis had collected, including the *Mahonia*. He went to England and was employed by A. B. Lambert, an avid amateur botanist. Lewis's plants became part of Lambert's collection, which, following Lambert's death, was auctioned off in 1840. Fortunately for American botany, an American in the audience appreciated the value of the Lewis plants, a small, seemingly insignificant part of Lambert's collection. The man, Edward Tuckerman, brought the plants back to the United States, where they were eventually rejoined with the rest of the collection in Philadelphia. Thus *Mahonia aquifolium* is among a select group that were carried across the country from the Pacific Coast to Philadelphia, then traveled to England and back, and are still in good condition today.

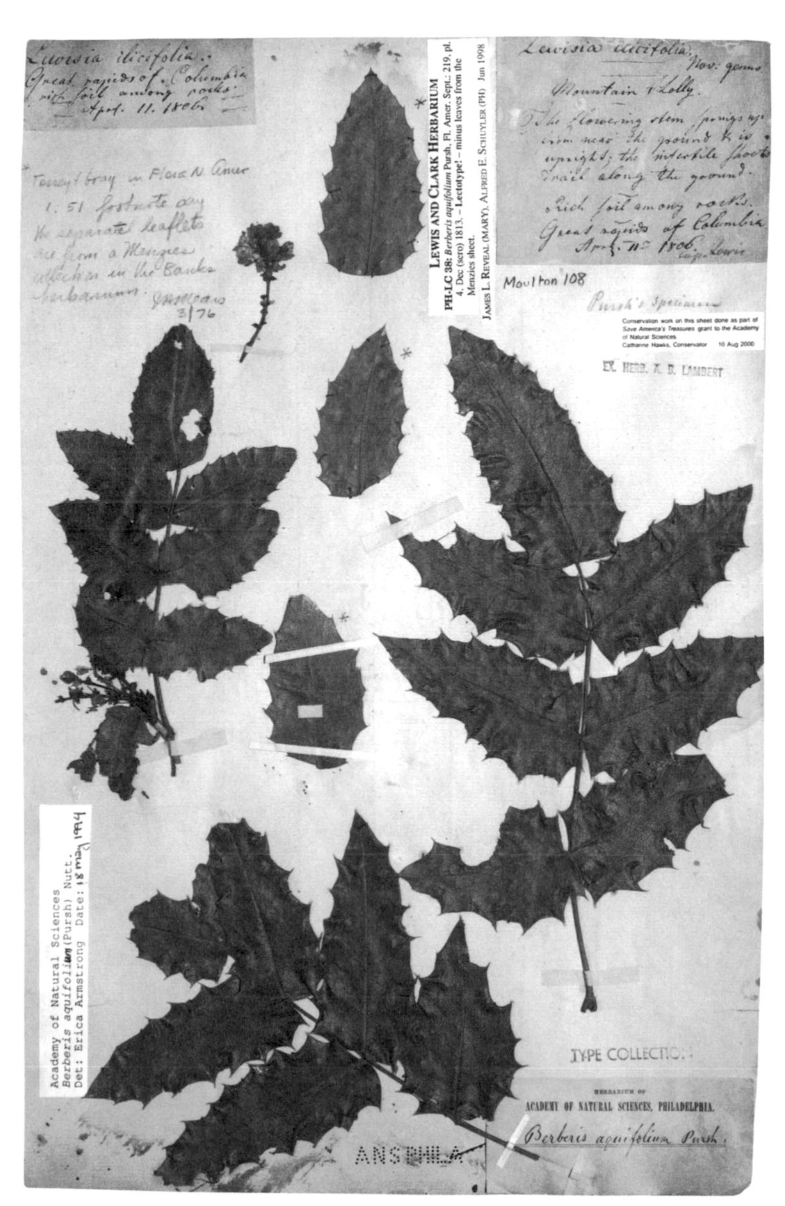

DURING THE WINTER OF 1806 LEWIS ACCURATELY DESCRIBED THE LEAF OF OREGON GRAPE HOLLY, *MAHONIA AQUIFOLIUM*, WITH ITS THREE PAIRS OF LEAFLETS EACH ARMED WITH THORNS AND SITTING RIGHT ON THE STEM. THE EVERGREEN LEAVES AND GRAPELIKE CLUSTERS OF FRUIT MAKE THIS AN ATTRACTIVE GARDEN SHRUB.

LEWIS'S SYRINGA

Philadelphus lewisii

Lewis's Syringa

PHILADELPHUS LEWISII

ON MARCH 23, 1806, the Corps of Discovery left its winter quarters at Fort Clatsop, on the Oregon coast, and began the journey homeward. It was a rainy, windy day, following a winter of such weather. The explorers were heading upriver once again, contending with the high water and swift currents of mountain snowmelt. Some of the men struggled with frequent portages while others went on extended hunting forays. As they made their way east the travelers had numerous encounters with American Indians; these events were sometimes friendly and sometimes edgy. The expedition hoped to make it over the Bitterroot and Rocky mountains and back down the Missouri River to St. Louis before the next winter.

In the coming months Lewis would be busy collecting new plant specimens. In April alone he collected some thirty flowering plants, including new species of trillium, larkspur, clover, and fritillaria. He also noted black hawthorn, thimbleberry, and other shrubs and trees. On the last day of April the expedition had just left the Columbia River for the Walla Walla River, and Lewis observed the changing vegetation.

> *there is a good store of timber on this creek at least 20 fold more than on the Columbia river itself. it consists of Cottonwood, birch, the crimson haw, red willow, sweetwillow, chokecherry, yellow currants, goosberry, whiteberryed honeysuckle rose bushes, seven bark, and shoemate. I observed the corngrass and rushes in some parts of the bottom.*

The expedition returned to a Walla Walla Indian village, where they had had a friendly encounter the previous fall. The chief, Yelleppit, presented Clark with an elegant white horse and provided the expedition with canoes. More important, the Indians gave them valuable information. They told the explorers about a shortcut to the Nez Perce Indians, eliminating about eighty miles of difficult terrain. A Shoshone woman living among the Walla Wallas talked with Sacagawea, and the two were helpful in answering questions the Walla Wallas had about the expedition. The discussion must have gone slowly because of the need to translate from the Walla Walla language into Shoshone and then into English and back again. But the end result was that the explorers "fully satisfyed all [the Walla Wallas'] enquiries with rispect to ourselves and the objects of our pursuit."

On May 6 Lewis collected a specimen of a shrub now known as Lewis's syringa. Other common names for *Philadelphus lewisii* are mock orange and Indian arrowwood. It has fragrant white flowers and its leaves are opposite. The stems of early growth are smooth and straight; on older stems the bark may be cracked or exfoliating. Indians used the wood for netting and to make needles and combs. Lewis collected another specimen of syringa on July 4 on the eastern side of the Rockies, this time without flowers. He made no comment on it at either time.

Several features of Lewis's syringa have led to speculation that the explorers might have seen it other times as well without connecting it to the collected specimens. Clark noted a plant that he thought might be a species of privet, a familiar garden shrub. Like *Philadelphus,* privet has small opposite leaves that are not toothed. Privet, however, is native to the Old World and could not possibly have been found in the Far West at that time. Another shrub, which Lewis referred to several times, is sevenbark, a wild hydrangea native to the East Coast. It, too, has exfoliating bark and opposite leaves and is closely related to *Philadelphus.*

We can be thankful to Frederick Pursh for selecting a scientific name for this western native that honors Meriwether Lewis. The common name, on the other hand, is cause for confusion. *Syringa* is the scientific name for lilac, and although there are some apparent similarities, taxonomically lilac and mock orange are not even close. Still, "syringa" has been applied to mock orange for centuries. The name suggests a hollow tube, and the story goes that the pith was removed from the stem, making it into a pipe and earning *Philadelphus* the name "pipe privet."

By May 6, when Lewis first collected his syringa, the expedition had reached the Clearwater River. The next day, in the vicinity of where he collected this plant, he wrote about his surroundings.

> *the face of the country when you have once ascended the river hills is perfectly level and partially covered with the longleafed pine. the soil is a dark rich loam thickly covered with grass and herbatious plants which afford a delightful pasture for horses. in short it is a beautifull fertile and picteresque country.*

Lewis's syringa is native to the Northwest and is adaptable, occurring in moist and dry, sunny and shaded situations. Hybrids of mock orange can be obtained to fit various landscaping needs. The native species *Philadelphus lewisii* is found in the wild, in native plant collections, and in occasional garden centers and nurseries.

GLACIER LILY
Erythronium grandoflorum

Glacier Lily

ERYTHRONIUM GRANDIFLORUM

ON MAY 8, 1806, the expedition had just crossed the border of present-day Washington and Idaho and were involved in intense discussions with the Nez Perce about obtaining horses. Food was another subject of concern; they were relying heavily on native plants to supplement their diet of meat. Lewis noted that the Indians used many plants, unfamiliar to him, for food. One was a root known as cous or cows, which resembles the highly poisonous water hemlock. He was afraid that his men, thinking they were gathering cous, might make a mistake and end up poisoning themselves, so he preferred to obtain it through trade with the Indians.

The explorers ascended more than a thousand feet between May 8 and June 15 as they approached the high mountains. Lewis collected specimens of glacier lily on both dates and made note of it on several other occasions, though he did not write a detailed description of the lily.

Glacier lily begins its bloom early, when temperatures rise and the snow melts, and the name reflects its habit of often appearing at the edge of snowbanks. Lewis referred to the plant as dogtooth violet, a familiar species of the eastern woodlands. He first collected it at an elevation where spring was well under way. The little yellow flower must have accompanied the Corps much of the way up the slopes, along with the advancing spring. The bulbs, leaves, flowers, and seed pods all provided food for the Indians, as well as for animals from small rodents to grizzly bears.

By June 15, the explorers were closer to the Bitterroot Mountains, which they were determined to cross as soon as possible. Over the next two days they ran into deeper and deeper snow; their horses were struggling and had no grass to feed on. Nevertheless Lewis found time to make careful ecological observations.

> *we saw in the hollows and N. hillsides large quatities of snow yet undisolved; in some places it was from two to three feet deep. vegetation is proportionably backward; the dogtooth violet is just in blume, the honeysuckle, huckburry and a small speceis of white maple are beginning to put fourth their leaves; these appearances in this comparatively low region augers but unfavourably with rispect to the practibility of passing the mountains.*

The captains reluctantly decided on June 17 to retreat for a while: "We conceived it madnes in this stage of the expedition to proceed without a guide."

Toward the end of June, advancing again, Lewis had words of praise for the Nez Perces, who were now guiding them.

> *from this place we had an extensive view of these stupendous mountains principally covered with snow like that on which we stood; we were entirely surrounded by those mountains from which to one unacquainted with them it would have seemed impossible ever to have escaped; in short without the assistance of our guides I doubt much whether we who had once passed them could find our way to Travellers rest...these fellows are most admireable pilots.*

He ended his entry with another botanical comment.

> *neare our encampment we saw a great number of the yellow lilly with reflected petals in blume; this plant was just as forward here at this time as it was in the plains on the 10th of may.*

The yellow lily referred to is the glacier lily, which, like its relative in the east, has six petals that curve back. The dogtooth violet of the eastern woodlands is also known as trout lily, and the name provides a clue to an obvious difference between these two related plants. The trout lily has mottled foliage, reminding many of the mottled sides of a trout. The glacier lily does not have mottled foliage, which no doubt Lewis observed.

Lewis collected another ephemeral bloomer of early spring, the yellowbell, *Fritillaria pudica,* which does not have reflected petals but otherwise is quite similar to the glacier lily. It, too, was a food source for local inhabitants. Viable seeds or bulbs of the yellowbell seem to have made it back to the United States. In April 1807 Thomas Jefferson noted in his Garden Book that he planted, among other things, "Lilly. The yellow of the Columbia. It's root a food of the natives."

Glacier lily is native to the western mountains, from the valleys to the timberline. It blossoms from April to midsummer, the time depending on altitude, and can carpet large areas with bright yellow flowers. It is adapted to take advantage of abundant moisture from melting snow and warm daytime temperatures of spring. This is a period of intense growth, when it stores energy in the bulb to tide it over the coming dry and then cold seasons. These beauties of the mountains are becoming rare and should be allowed to flourish in their native habitat.

RAGGED ROBIN
Clarkia pulchella

Ragged Robin

CLARKIA PULCHELLA

ON MAY 14 THE EXPLORERS arrived at a spot on the edge of the Clearwater River in present-day Idaho that has come to be known as Camp Chopunnish, for the name by which they referred to the Nez Perce Indians. They were to spend almost four weeks here, occupied with laying in provisions, trading and socializing with the Nez Perce, gelding horses, and tending to a variety of health problems among themselves and the Indians. During this time Lewis observed a number of plants, writing detailed descriptions of several and commenting on the potential commercial value of others. It was a safe and pleasant spot, and "as we are compelled to reside a while in this neighbourhood I feel perfectly satisfyed with our position."

On June 1 Lewis wrote about ragged robin in one of his most detailed botanical descriptions. This is another plant that would later be formally named by Frederick Pursh, and this time he chose to honor Captain William Clark. *Clarkia* is a genus of plants native to western North America. Ragged robin is also sometimes known as beautiful Clarkia. Another plant native to the eastern part of the continent and also called ragged robin is not related and is much more raggedy.

Lewis pulled out all the stops when he wrote about the western ragged robin, using numerous technical botanical terms. He clearly was enchanted by this flower and ended his entry with the words "I regret very much that the seed of this plant are not yet ripe and it is proble will not be so during my residence in this neighbourhood." He began his account as follows.

> *I met with a singular plant today in blume of which I preserved a specemine; it grows on the steep sides of the fertile hills near this place, the radix is fibrous, not much branched, annual, woody, white and nearly smooth. the stem is simple branching ascending, 2½ feet high celindric, villose and of a pale red colour....*

The word *radix,* not much used today, refers to the root. *Villous,* or *villose,* as Lewis wrote it, refers to long, soft hairs. He noted that the leaves and the pistil are also villous and that the leaves are thinly scattered and sit right on the stem.

Lewis gave more details on the appearance of the leaves and then described the flower.

> *the calyx is a one flowered spathe. the corolla superior consists of four pale perple petals which are tripartite, the central lobe largest and all terminate obtusely; they are inserted with a long and narrow claw on the top of the germ, are long, smooth, & deciduous.*

Flowers of ragged robin can range from a lavender so pale that it is almost white to pale purple; perhaps the ones Lewis saw were uniformly purple. He then went on to describe the stamens.

> *there are two distinct sets of stamens the 1st or principal consist of four, the filaments of which are capillary...; the anthers are also four...are linear and reather flat, erect sessile,...twise as long as the fillament....the second set of stamens are very minute are also four and placed within and opposite to the petals, these are scarcely persceptable....the anthers are four, oblong, beaked, erect...and appear not to form pollen.*

Lastly he described the pistil with its stigma bearing four white lobes. The four-lobed stigma is a characteristic common to the primrose family, to which *Clarkia pulchella* belongs.

> *the single style and stigma form a perfict monapetallous corolla only with this difference, that the style which elivates the stigma or limb is not a tube but solid tho' its outer appearance is that of the tube of a monopetallous corolla swelling as it ascends and gliding in such manner into the limb that it cannot be said where the style ends, or the stigma begins; jointly they are as long as the corolla, white, the limb is four cleft, sauser shaped, and the margins of the lobes entire and rounded. this has the appearance of a monopetallous flower growing from the center of a four petalled corollar, which is rendered more conspicuous in consequence of the 1st being white and the latter a pale perple.*

Reading Lewis's observations of ragged robin, no one can say that he wasn't a natural-born botanist. He clearly took delight in examining closely this delicate and unfamiliar plant and searching for the right word to describe its each and every identifying characteristic. His description is one to send even the expert running to the dictionary of botanical terms.

Among the various notes and letters written by the men of science in Philadelphia, there is nothing to indicate that Lewis brought back seed of ragged robin for them to germinate. Later botanists going west did, however, bring it back and it was evidently grown in gardens by the 1820s. Other species of *Clarkia* go by colorful names such as farewell-to-spring and elegant Clarkia. Elegant Clarkia and beautiful Clarkia are annuals that are easy to grow from seed, are effective when planted in a mass, and hold up well as cut flowers if gathered while still in bud.

SILKY LUPINE
Lupinus sericeus

Silky Lupine

LUPINUS SERICEUS

THE WESTERN MOUNTAINS of North America are home to many species of lupine. Lewis encountered at least four. He wrote about one during the winter of 1805–06 on the Pacific coast and collected three others while in the mountains on the trip home. On June 5, 1806, while at Camp Chopunnish in present-day Idaho, Lewis made note of a considerable number of plants that he thought were "common to our country," including "several of the pea blume flowering plants." This is thought to be a reference to the silky lupine, which is not found in the eastern part of the continent; it was unknown to western science. The flower of silky lupine is of a pale lavender hue. It is typical of the pea family, with its distinctly different upper and lower petals. The Lewis and Clark Herbarium of the Academy of Natural Sciences houses specimens of the plant, which he collected on this date and again on July 7 near Lewis and Clark Pass in Montana.

In the winter both captains had become familiar with a species of lupine that grows near the seacoast and has a root that tastes like liquorice. This lupine (*L. littoralis*) was a staple in the diet of the Chinook and other coastal Indian tribes. Members of the Corps of Discovery were also glad to eat it. Lewis did not realize this also was a plant new to western science. He did not collect it but did write about it.

I observe no difference between the liquorice of this country and that common to many parts of the United states where it is also sometimes cultivated in our gardens. this plant delights in a deep loose sandy soil; here it grows very abundant and large; the natives roast it in the embers and pound it slightly with a small stick in order to make it seperate more readily from the strong liggament which forms the center of the root....this root when roasted possesses an agreeable flavour not unlike the sweet pittaitoe.

On the return home, during the Corps' long stay at Camp Chopunnish, Lewis collected sixteen plant specimens. But this was not his only responsibility. Both captains tended to various health problems. They utilized the local flora when their medical supplies failed to do the job. William Bratton, a member of the expedition, had become so weak in the small of his back that he could hardly walk. A much respected Indian chief had lost the use of his limbs and was brought to the camp for help. The captains subjected Bratton to a kind of sauna, where he sat for a number of hours sweating profusely and drinking strong "horsemint" tea. He immediately felt better, so they recommended the same treatment for the chief. Without knowing the true identity of the horsemint, we cannot say what effect it had on the invalids. Fortunately they also had plenty of pure, cold mountain water to drink. Sacagawea's son, Pomp, had been suffering from a swollen neck and jaw. The remedies they tried for him included a poultice of onions. By early June, somewhat astonishingly, all three patients were recovering nicely. Near the end of their stay, the travelers experienced a health problem of an altogether different kind. Lewis observed that those men "who are not hunters have had so little to do that they are geting reather lazy and slouthfull." To get back in shape the men ran foot races with the Indians and played "prison base."

After the expedition left Camp Chopunnish, Lewis collected two other new species of lupine. They are silvery lupine, *Lupinus argenteus*, and rusty lupine, *Lupinus pusillus*, both western natives that were new to science.

There are two specimens of lupine in the Herbarium; three more thought to have been collected by Lewis are in the Royal Botanic Gardens at Kew in England. How did three of them end up at Kew Gardens in England bearing the stamp of the Hooker Herbarium? Once again Frederick Pursh enters the story.

When Pursh left Philadelphia for England in late 1811 with some of the Lewis and Clark specimens, he apparently took along some of the lupine. These plants came into the possession of Aylmer Bourke Lambert, who employed Pursh. Lambert, a respected amateur botanist, for a time acted as director of the gardens at Kew. The gardens were very popular among the British public but did not have formal status as a national botanic garden. A certain lord, not appreciating the value of the already considerable Kew plant collections, tried to dismantle them and convert one of the plant houses into a vinery. Thanks to the foreman and a kitchen gardener, word got out, the public went into an uproar, and the gardens and collections were saved. The great botanist William Hooker was appointed director, and under his guidance Kew Gardens began to flourish. About forty of the Lewis and Clark plants, in Lambert's collection thanks to Pursh, were taken back to the United States, but a few were moved to Kew. Today some eleven of the specimens at Kew are thought to have been collected by Lewis.

The name *lupine* comes from the Latin for wolf. At one time lupines were thought to rob the soil of minerals; instead, being a member of the legume family, they have the ability to fix nitrogen, thus improving soil fertility. Many lupines are poisonous, but others are edible.

The three species that Lewis collected range from lavender to blue to purple. Other lupines bloom in shades of pink, white, and yellow. Modern hybrids come in even brighter and bolder colors. Lupines are popular as garden flowers because of their 12-inch-tall flower spikes and deeply cut palmate foliage. Most are perennials.

OLD MAN'S WHISKERS
Geum triflorum

Old Man's Whiskers

GEUM TRIFLORUM

GEUM IS A GENUS OF PLANTS found throughout much of the northern hemisphere. Some species are native to mountainous regions, including the Himalayas, Caucasus, and Pyrenees. Lewis collected a specimen of *Geum triflorum,* old man's whiskers, on June 12, 1806, while the expedition was camped at Weippe Prairie, at the western edge of the Bitterroot Mountains. They had just departed Camp Chopunnish.

He made no comment in his journal about old man's whiskers or another newly found plant that he collected the same day, the western snakeweed (*Polygonum bistortoides*). An annotation on the Herbarium sheet to which the snakeweed is attached says it was found "in moist ground on Quamash flats." Lewis did comment, however, on a plant that had become very familiar to him, the camas (quamash), which was in bloom and from a distance resembled a lake of clear water.

Like camas, *Geum triflorum* blooms with the melting snows of spring and early summer. It is found in rocky places and prairies where moisture is not constant year round. Its nodding pink flowers usually appear in threes, as the name *triflorum* indicates, perched atop a stem that is also pinkish. As the flower matures and goes to seed the stems turn upward and the styles lengthen up to two inches and become feathery. The styles are part of the female organs of the plant that connect the stigma, where the male pollen grains land, with the ovary. Each style is attached to a seed. After the seed has been fertilized, the feathery styles are caught by the wind and carry the seed away. The styles curve in opposite directions, a feature that may further ensure seed dispersal over a wide area and increase the plant's chances for success in producing a new generation.

These unusual, feathery styles account for the imaginative common names for this plant, including old man's whiskers, prairie smoke, tassels, long-plumed avens, and pink plumes. Lewis collected old man's whiskers when it was still in flower, in other words, before it had become whiskery, tassely, plumy, or in any way resembled smoke. As the season advanced and the expedition left the mountains for the prairies, the geum would have been going to seed and, although Lewis never commented on it, he might have seen it in its plumy appearance.

For months Lewis had been collecting plants native to the Northwest. Now he had collected one whose range extended east to western New York state. *Geum triflorum* is a northern plant and does not flourish south to places like Virginia and Georgia, where Lewis had rambled in much of his youth. None of the botanists exploring in eastern North America had formally described it, and so it remained for Lewis to collect it in the Bitterroot Mountains and for Frederick Pursh to name it for science.

Several days after collecting the geum, Lewis wrote about the mood of the expedition, which was one of eagerness to get home. The men had packed up everything and hobbled the horses to prevent their straying during the night. In the morning they set out for Travellers Rest.

> *we have now been detained near five weeks in consequence of the snows; a serious loss of time at this delightfull season for travelling. I am still apprehensive that the snow and the want of food for our horses will prove a serious imbarrassment to us as at least four days journey of our rout in these mountains lies over hights and along a ledge of mountains never intirely destitute of snow. every body seems anxious to be in motion, convinced that we have not now any time to delay if the calculation is to reach the United States this season; this I am detimined to accomplish if within the compass of human power.*

Clark's entry for the same day speaks of "those Snowey tremendious mountains" which "I Shudder with the expectation" of passing over. They soon ran into deep snow and had to retreat.

On June 25 they set out again. The terrain was extremely rugged and in places the snow was still seven feet deep. Thanks to their guides, the party made it to Travelers Rest the same day. On June 30 Lewis's horse slipped, Lewis fell some forty feet, and the horse almost landed on top of him. Fortunately both escaped injury. Lewis noted the fall in his journal and then described an unknown plant, the beautiful and rare mountain lady slipper.

The genus *Geum*, which is also known as avens, is a member of the rose family. A typical characteristic of this family is to have many sexual parts per flower, as seen in the numerous styles of old man's whiskers. Geums are cold-hardy perennials with deeply cut basal foliage. The flowers come in various colors, including white, yellow, orange, scarlet, and red. A number of hybrids are on the market and can be purchased in garden centers and through catalogs. *Geum triflorum* is available from suppliers that specialize in native plants of the prairie.

SHRUBBY PENSTEMON
Penstemon fruticosus

Shrubby Penstemon

PENSTEMON FRUTICOSA

BY THE SUMMER of 1806 a year had passed since Lewis's shipment from Fort Mandan had arrived in the United States. Many at home assumed the worst had happened to the Corps of Discovery. However, the month of June was the busiest of the entire expedition in terms of Lewis's plant collecting. Thirty-nine specimens in the Herbarium date from this month alone. All the plants that he had collected since the preceding fall were pressed, dried, and safely stored away. Considering the dampness of the winter, frequent swampings of canoes on turbulent rivers, and packages falling off horses when the animals slipped on snowy tracks, it is remarkable that these botanical treasures survived.

In May, while the party was still on the Clearwater River, Lewis collected Cascade penstemon (*Penstemon wilcoxii*), a plant that is somewhat taller than shrubby penstemon and with a bluer flower. He found the specimen of shrubby penstemon in mid-June, in the vicinity where he also collected old man's whiskers, glacier lily, a species of trillium, beargrass, and a number of other new plants. He noted that much of the country was covered with pine, cedar, arborvitae, and fir. Throughout the expedition, he also made numerous observations of animals. On June 15 he noted a specked woodpecker, a bee martin, a large cock, and freshly laid eggs in a hummingbird's nest.

Although spring had come to the mountains, deep snow remained in many places, holding back the forward movement of the expedition. The shrubby penstemon was in bloom benefiting from abundant water provided by melting snow. By July its flowers would be finished; it would dry out in the rocky soil for the rest of the summer and fall and eventually be protected from foragers and the elements by a blanket of snow.

Shrubby penstemon is one of several species of penstemon native to the western mountains of North America. A typical feature of the genus is the bilaterally symmetrical flower that is composed of two matching halves rather than showing the same symmetry all the way around (as on the daisy, for example). Moving on from genus to family, penstemons are grouped with other familiar bilaterally symmetrical garden flowers, including snapdragon, turtlehead, foxglove, mimulus, and veronica. Indian paintbrush is also a relative.

The defining feature of penstemons is the presence of five stamens, which also explains the origin of the name (*penta-*, from the Greek, means five). The pollen-containing anthers are part of the stamen, the male sexual organ of a flower. In the penstemons, the fifth stamen has no anther and therefore no pollen, which means it is sterile. To be fully accurate, the fifth stamen isn't a stamen at all, but a stamenode. It is often showier than the four true stamens.

The lavender flowers of shrubby penstemon are long and tubular. The wide lower lip provides a landing platform that allows pollinating insects to crawl inside and reach the nectar. Other penstemons do not offer such an amenity and are pollinated by hummingbirds, which hover outside the flower and reach the nectar with their long beak.

When Lewis collected his two specimens of penstemon, they were in bloom. If he'd been a little later and collected seeds to take east, they might have found their way into the hands of Bernard McMahon. McMahon and Jefferson corresponded for a number of years after the return of the Corps of Discovery and mentioned Lewis's botanical discoveries on various occasions. In 1807 Jefferson sent some of Lewis's seeds to McMahon to propagate. Because he was still president, Jefferson wrote that he was "not in a situation to do them justice." He added a curious note. He suggested that when McMahon next saw Lewis he not tell him he had already received seeds from Jefferson. This way Lewis would be inclined to hand over more seeds still in his possession. McMahon told Jefferson he had never seen seeds so well preserved as those collected by Lewis. He hoped Lewis would accept a gift from him, a collection of culinary and ornamental plants, which he would send whenever Lewis was ready for them.

Penstemon, or beard tongue, as it is also called, is largely a North American genus. Eastern species are somewhat more tolerant of moist conditions throughout the year; western species need long periods without rain. Essentially, this group of plants likes dry situations and is appropriate for rock gardens; it should not be planted in humus-rich, moisture-retaining soil. Seeds of many varieties are available.

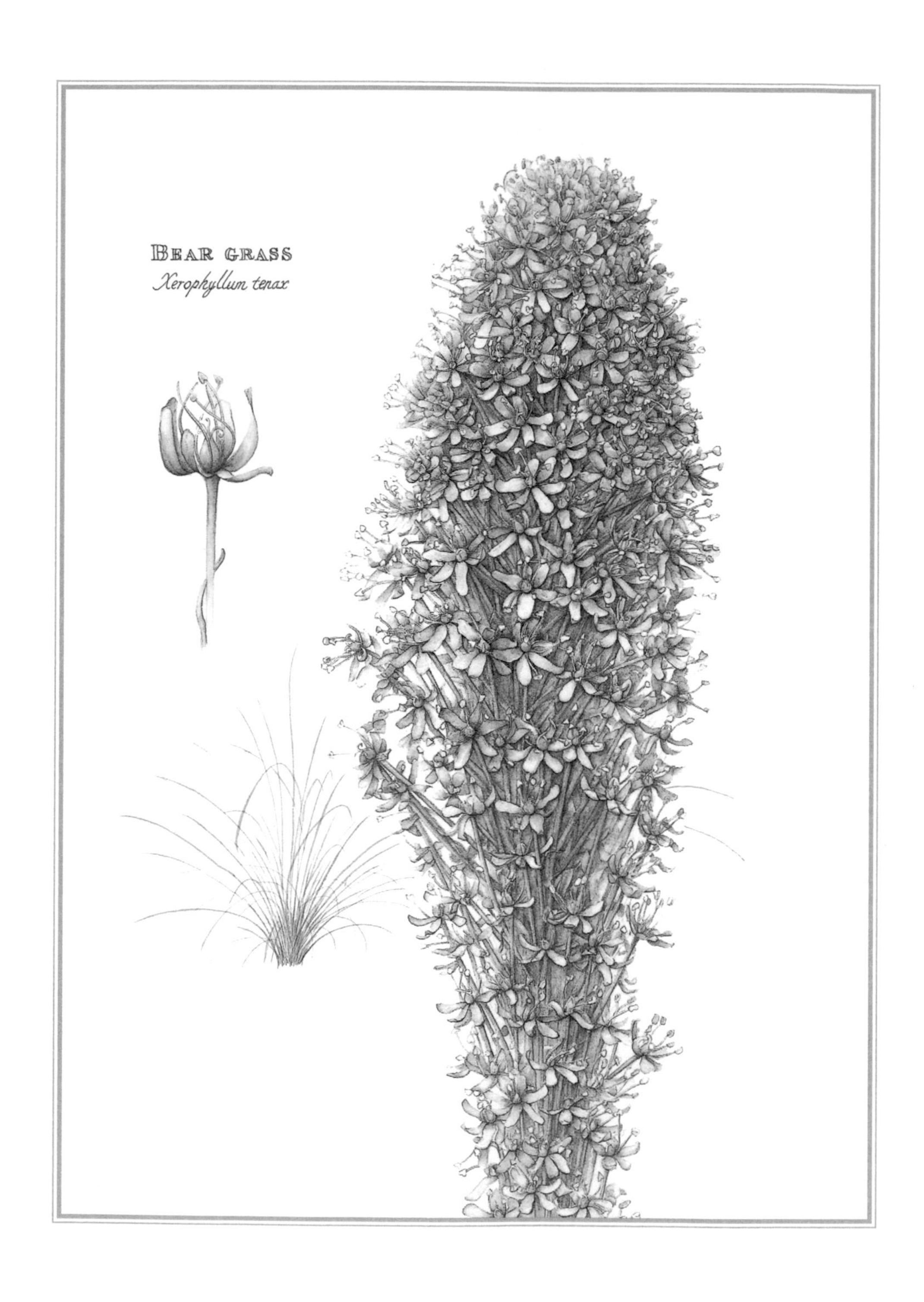
Bear grass
Xerophyllum tenax

Bear Grass

XEROPHYLLUM TENAX

IN MID-JUNE 1806, when Lewis collected this unusual flower in the Bitterroot Mountains, he made no note of it; he had become familiar with the plant long before he saw it in bloom. Not a grass at all but a member of the lily family, bear grass is native to the Northwest from near sea level to elevations of 7,000 feet. Its tough evergreen leaves grow close to the ground. In the mountains, it must survive dry winds and blowing ice crystals.

From May to late summer bear grass puts up a stalk three to four feet tall topped with a club-shape cluster of flowers. Each individual flower is only about half an inch across. The roots, leaves, flowers, and seeds provide food for many animals. Mountain goats eat the tough leaves in winter; bears eat the tender parts of the leaves in spring. Bear grass does not bloom every year; in fact, as many as ten years may elapse between flowerings. In years when many plants bloom at once, they make a beautiful sight, filling high mountain meadows with erupting dense clusters of creamy white flowers set against of a backdrop of snow-covered peaks.

Both Lewis and Clark recorded bear grass a number of times, starting in the fall of 1805. Several times they saw Indians going downriver in canoes loaded with bear grass and other items of trade. After the expedition had settled at Fort Clatsop, on the Oregon coast, the captains wrote about the conical hats the Indians wore and made little sketches of them in their journals. They were quite taken with these hats and described them in some detail.

> *they wear a hat of a conic figure without a brim confined on the head by means of a string which passes under the chin and is attatched to the two opsite sides of a secondary rim within the hat. the hat at top terminates in a pointed knob of a connic form also....these hats are made of the bark of cedar and beargrass wrought with the fingers so closely that it casts the rain most effectually....on these hats they work various figures...faint representations of whales the canoes and the harpoonneers striking them. sometimes squares dimonds triangles &c.*

The captains had their head measurements taken and ordered hats to be made for themselves, which turned out to fit very well. They purchased hats for the rest of the expedition. Given that it had rained so much of the winter, the headgear must have been especially welcome. Lewis admired the exceptional skill of the Indians and took one of the hats back to the East Coast.

The Indians also used bear grass to make baskets. Thus another name for it is Indian basket grass.

> *their baskets are formed of cedar bark and beargrass so closely interwoven with the fingers that they are watertight without the aid of gum or rosin; some of these are highly ornamented with strans of beargrass which they dye of several colours and interweave in a great variety of figures; this serves them the double purpose of holding their water or wearing on their heads; and are of different capacites from that of the smallest cup to five or six gallons....these they make very expediciously and dispose off for a mear trifle.*

Lewis collected his specimen on June 15, a rainy day when the horses frequently stumbled on the trail. Bear grass leaves, which are exceptionally slippery when wet, no doubt contributed to their difficulties. Eleven days later he wrote about its abundant growth.

late in the evening...we arrived at the desired spot and encamped on the steep side of a mountain....here we found an abundance of fine grass for our horses. this situation was the side of an untimbered mountain with a fair southern aspect where the snows from appearance had been desolved about 10 days.... there is a great abundance of a speceis of bear-grass which grows on every part of these mountains it's growth is luxouriant and continues green all winter but the horses will not eat it.

The name *Xerophyllum tenax* is apt; it means tough, dry leaves. *Xerophyllum* is a North American genus of only two species, bear grass in the west and turkey beard (*X. asphodeloides*) in the east, which is found from the New Jersey pine barrens south to Georgia. How Lewis and Clark came to call the western species bear grass is not known.

Bear grass is not a plant for the garden, since it may not bloom for years at a time and it is difficult to grow. This is a plant best enjoyed in its native habitat.

PONDEROSA PINE
Pinus ponderosa

Ponderosa Pine

PINUS PONDEROSA

IN THE FALL OF 1805 and spring of 1806 the expedition often saw ponderosa pine, or the "longleafed" pine, as Lewis and Clark called it in their journals. They noted it throughout September 1805 as the expedition struggled through the Bitterroot Mountains, and on October 1 Lewis collected some of the long leaves, or needles.

The following spring, retracing their route up the Columbia River, they began to see ponderosa pine again as they advanced inland. In April they toiled their way along the Columbia through the Cascade Mountains and then entered the higher and drier central plateau region of present-day Oregon and Washington. This is the western edge of the range for ponderosa pine, which is predominantly an inland species. On April 14 Lewis wrote, "near the border of the river I observed today the long leafed pine. this pine increases in quantity as you ascend the river and about the sepulchre rock where the lower country commences it superceedes the fir altogether." Ponderosa pine thrives in full sun in a dry climate where the soil is moist and fertile. The same conditions are good for grasses, and on more than one occasion, when the party camped near ponderosa pine, they were able to feed their horses well.

The Indians valued the ponderosa pine as a source of food in hard times. They ate the succulent inner bark, which they knew how to remove without killing the tree. The explorers were told that the previous winter had been a bad one, and on May 8 Lewis wrote

I observed many pine trees which appear to have been cut down...in order to collect the seed of the longleafed pine which in those moments of distress also furnishes an article of food; the seed of this speceis of pine is about the size and much the shape of the seed of the large sunflower; they are nutricious and not unpleasant when roasted or boiled.

The seeds of the ponderosa pine are a favorite food of squirrels, chipmunks, and birds. While at Camp Chopunnish, Lewis killed several birds that he had observed for some time and thought to be a species of crow. He described it at length, including its call, size and shape, color of its feathers, and number of toes, and concluded with "this bird feeds on the seed of the pine and also on insects. it resides in the rocky mountains at all seasons." The bird, which was new to science, is known today as Clark's nutcracker. It is often seen in ponderosa pine, using its powerful beak to get at the seeds. During the course of the expedition Lewis described some fifty birds. Another, which he saw along the Missouri River, is known as Lewis's woodpecker.

Indians of the American Northwest not only ate the seeds and bark of ponderosa pine, they also chewed the rosin and used it as a salve and for waterproofing. Among the treatments that Captain Clark used when he tended Sacagawea's sick baby's neck at Camp Chopunnish, he applied a salve made of rosin of long-leaf pine mixed with beeswax and bear oil to Pomp's neck. As already mentioned, the child recovered.

In late June the Nez Perce, who were guiding the expedition, provided an evening of entertainment. They set fire to the trees. The sudden and immense blaze reminded Lewis of a fireworks display. Like a fireworks display, it was over quickly and caused little damage. The Indians explained that the fire would bring good weather to the explorers for the rest of their journey. Fires like this, whether set by lightning or by the inhabitants, did not damage the ponderosa pine. Its thick bark enables the mature tree to withstand fire. Sometimes pure (single-species) stands of ponderosa pine are seen; these are thought to be the result of fire that has killed less resistant species and left only the ponderosa.

By the first of July the expedition had crossed over the most difficult part of the mountains, and they would soon part company with their guides. They recorded seeing ponderosa pine this time associated with other trees and shrubs.

> *the wild rose, servise berry, white berryed honeysuckle, seven bark, elder, alder aspin, choke cherry and the broad and narrow leafed willow are natives of this valley. The long leafed pine forms the principal timber of the neighbourhood, and grows well in the river bottoms as on the hill. the firs and larch are confined to the higher parts of the hills and mountains.*

The range of ponderosa pine is throughout the western mountains and, in lesser numbers, east across the northern plains to North Dakota. The bark is yellowish to cinnamon red and the needles are a yellowish green. Another name for it is western yellow pine. The needles, usually in bundles of three, are four to eleven inches long. Ponderosa pine may live as long as 600 years and reach 200 feet in height. When the Corps of Discovery passed through this country, before the days of massive human impact on the environment, many of these trees had achieved tremendous growth. Today ponderosa pine is important commercially for its timber.

BITTERROOT
Lewisia rediviva

Bitterroot

LEWISIA REDIVIVA

THIS SMALL SUCCULENT alpine plant gave its name to a range of enormous mountains and a river; it is the state flower of Montana and was named for its discoverer, Meriwether Lewis. Lewis hardly made note of it in his journals, yet this plant is one of the most widely recognized in connection with the Corps of Discovery.

On July 2, 1806, the expedition was at Travellers Rest in the Rocky Mountains, near present-day Missoula, Montana. Lewis noted seeing some seventeen plants, though he wrote that nothing worth mentioning had happened.

> *I found two speceis of native clover here, the one with a very narrow small leaf and a pale red flower, the other nearly as luxouriant as our red clover with a white flower the leaf and blume of the latter are proportionably large. I found several other uncommon plants specemines of which I preserved.*

One of these uncommon plants was the bitterroot. The pressed and dried flowers still exist in the Lewis and Clark Herbarium; the roots, which Lewis also collected, were eventually planted by Bernard McMahon.

It seems unlikely that Lewis realized the flowering plant was the same one whose bitter roots he had tried eating a year earlier. He had first encountered bitterroot In 1805 as the expedition, westward bound, approached the Continental Divide. One August day he tried three different kinds of roots, which he had learned from the Indians were edible. Two he found quite agreeable and the third, the bitterroot, he did not enjoy.

another speceis was much mutilated but appeared to be fibrous; the parts were brittle, hard of the size of a small quill, cilindric and as white as snow throughout, except some small parts of the hard black rind which they had not seperated in the preperation. this the Indians with me informed were always boiled for use. I made the exprement, found that they became perfectly soft by boiling, but had a very bitter taste, which was naucious to my pallate, and I transfered them to the Indians who had eat them heartily.

On the homeward journey in 1806, July 2 marked a significant moment for the expedition. During the previous winter the captains had developed a plan to split up when they reached this point. Each would take a small party and explore unknown regions, Lewis going to the north up the Marias River and Clark to the south along the Yellowstone River. They would rendezvous a month later at the junction of the Yellowstone and Missouri rivers. It must have been a moment of excitement but also of apprehension. The next day Lewis wrote

I took leave of my worthy friend and companion Capt. Clark and the party that accompanyed him. I could not avoid feeling much concern on this occasion although I hoped this seperation was only momentary.

It was at great risk that they undertook this additional exploration now that they had completed the expedition's major goal of reaching the Pacific Ocean, mapping unknown territory, collecting specimens, and keeping detailed records of everything they had seen. If they experienced disaster now, all they had learned since leaving Fort Mandan in the spring of 1805 would be lost. They did, of course, make the planned rendezvous in August and safely finish the journey home.

The roots of *Lewisia rediviva* that Lewis collected in the mountains also made it safely to the United States almost three months later. As further testament to their hardiness, they were then shipped to Washington, D.C., where Thomas Jefferson sent them to Bernard McMahon in

Philadelphia. McMahon planted the roots, which produced foliage but failed to bloom. This is not surprising given that McMahon, despite his great knowledge of plants, did not know how to grow an alpine plant native to the far side of the Rocky Mountains.

Frederick Pursh gave the plant its name. He recognized this member of the portulaca family as belonging to a genus new to science and he named it *Lewisia* after its discoverer. *Rediviva* refers to the root's ability to return to life following prolonged dormancy. Pursh reported seeing the plant that McMahon grew and, having seen the dried flowers, regretted that it never bloomed. In his work on the flora of North America, he wrote, it would be a desirable addition to an ornamental flower garden.

All members of the genus are alpine species native to the Pacific Northwest; some are evergreen, and others, deciduous. They have fleshy, succulent foliage and deep, cormlike tap roots suited to survival during dry periods. Lewis collected one other species of *Lewisia* about the same time *(Lewisia triphylla)*.

Lewisia rediviva is deciduous; the foliage appears in the spring but dies down when the flower appears. After flowering, the plant becomes dormant for the hot dry season, a required part of its life cycle. It grows in dry, gravely, well-drained soil. It is a small plant, only about two inches tall, with a basal rosette of narrow leaves about three inches long.

The flower is white to rose colored and is the most conspicuous element of the plant. It has twelve to eighteen waxy petals, measures up to four inches across, and has been described variously as sumptuous, improbable, unexpectedly pretty, and resembling a water lily. The flowers appear between April and July and then the plant goes dormant. Rock gardeners today who grow lewisias in wet climates need to provide protection against rain in order to get them to bloom. Modern hybrid lewisias, with larger blossoms and in more colors, are commercially available.

WOOD LILY
Lilium philadelphicum

Wood Lily

LILIUM PHILADELPHICUM

THE WOOD LILY IS a native plant of the upper Midwest. It is not to be found in the Lewis and Clark Herbarium, nor is there any mention of it in the journals, Frederick Pursh however, wrote that he saw specimens that had been collected by both Lewis and Thomas Nuttall, and that "the flowers are of an uniform deep scarlet colour, and are highly ornamental."

Wood lily blooms from June to August. Found mostly east of the Continental Divide, its habitat is wet and dry woods, glades and meadows, prairies, and shorelines. Lewis may have come across it in July of 1806 as the expedition was leaving the mountains and passing through moist meadows with rich, fertile soil crossed with creeks and streams. In early July he described a spot where the wood lily is likely to grow.

> *thus far a plain or untimbered country bordered the river which near the junction of these streams spread into a handsome level plain of no great extent; the hills were covered with long leafed pine and fir. I now continued my rout up the N. side of the Cokahlahishkit [Blackfoot] river through a timbered country for 8 miles and encamped in a handsom bottom on the river where there was an abundance of excelence grass for our horses.*

The deep orange-red flower is three to four inches across and stands erect on the stem, unlike similar species whose flowers nod. Dark purple spots are sprinkled around its throat. The base of the petals is markedly narrower than the blade. Usually there is just a single flower, less frequently several, on a stem.

There are eastern and western varieties of *Lilium philadelphicum*, which are very similar. The range of the western variety, the one Lewis would have seen at this time, extends across the northern Great Plains to the Rocky Mountains, from lower Canada southward to Nebraska. It is the official floral emblem of the province of Saskatchewan. Also called western orange-cup lily, prairie lily, or mountain wood lily, the plant grows one to two feet tall, not as tall as the eastern variety. It has one set of whorled leaves at the top of the stem and sparse foliage below; the eastern variety has several sets of whorled leaves along the stem.

The mountain wood lily was named by Thomas Nuttall, who according to Pursh had collected it near Fort Mandan, in North Dakota. Nuttall gave it the name *andinum*, "of the Andes," referring to the western cordillera of North and South America; apparently he was under the misapprehension that North Dakota was near the Rocky Mountains. There is no way of knowing whether Lewis or Nuttall were familiar with the eastern variety and recognized the distinction between the two.

Many years before Lewis saw the wood lily, a specimen of the eastern variety had been sent to Europe, where it was described by Linnaeus. Throughout the 18th century plant lovers on both sides of the Atlantic were sharing discoveries, shipping plants back and forth to one another. The extensive correspondence between two Quaker gentlemen sheds much light on which plants were most sought; John Bartram, a Pennsylvania farmer, was a self-educated plantsman. London cloth merchant Peter Collinson was an enthusiastic and well-connected gardener. His passion for plant collecting meshed perfectly with Bartram's innate ability. Their letters are a rich legacy. In 1730 Bartram sent Collinson wood lilies. A few years later, the Englishman requested that he send lilies of all sorts, as well as Solomon's seal, lady slipper, blazing star, and others. In return, Bartram asked for seeds of English wildflowers. The correspondence between Bartram and Collinson, who never met in person, lasted for thirty-five years until Collinson's death in 1768.

Many other North American seeds, plants, and flowers were sent to England; among these were hellebore, aster, laurel, Canada lily, American Turk's cap, and even skunk cabbage. Old World species of lilies, such as the madonna lily and the scarlet Turk's cap lily, were sent to America. Years later, Thomas Jefferson had both native and imported lilies in his garden, including the American and scarlet Turk's cap, yellow Canada, the madonna, and the blackberry lily.

Although the cut flowers do not last in a bouquet and, of course, can no longer produce seed, overpicking has endangered the mountain wood lily. With so many other lilies to choose from that are not at risk, this beautiful wildflower should be left to thrive in its native habitat.

YELLOW MONKEYFLOWER
Mimulus guttatus

Yellow Monkeyflower

MIMULUS GUTTATUS

ON JULY 3, 1806, Captain Clark rounded up his horses, said good-bye to his cocaptain, and set off with a party of twenty-three, Sacagawea among them, to explore the Yellowstone River. About ten days later Clark wrote that the Shoshone woman "has been of great Service to me as a pilot through this Country," the land where she had lived as a child.

When the party reached the Missouri they split up again. One group went down the Missouri to meet Lewis. The other, led by Clark, headed for the Yellowstone River, or Rochejhone as he referred to it, from the French *roche-jaune,* "yellow rock." The horses' feet became very sore on the gravelly ground and at one point Clark made moccasins to put over their hooves. After they had reached the Yellowstone one of the men injured himself and could no longer ride. Clark searched for several days to find trees, in that area of mostly cottonwood, that would be large enough to make into canoes. He finally had to be satisfied with two small canoes, which the men lashed together for increased stability. In these the party of nine, including Sacagawea and her son, Pomp, rode down the Yellowstone to the Missouri. They were troubled by mosquitoes and were sometimes kept awake at night by the bellow of mating buffalo, but otherwise carried out their exploration uneventfully. On August 3, after successfully completing the trip down the Yellowstone, Clark declared it a delightful river.

On the same day that Clark set out for the Yellowstone, Lewis and nine men headed north with their Nez Perce guides on an exploration that would turn out to be far more perilous than Clark's. By evening the Indians told Lewis they would go no farther. They were fearful of enemy tribes, which they were sure would give Lewis trouble as he proceeded east. The next day, July 4, he provided the Nez Perce with food for their journey back across the mountains, "smoked a pipe with these friendly people and at noon bid them adieu." He did not observe the national holiday, unlike Clark who wrote

> *This being the day of the decleration of Independence of the United States and a Day commonly Scelebrated by my Country I had every disposition to Selebrate this day and therefore halted early and partook of a Sumptious Dinner of a fat Saddle of Venison and Mush of Cows (roots).*

On July 4 Lewis collected yellow monkeyflower, perhaps along one of various creeks he noted in the area. *Guttatus* means spotted, referring to the dark red spots in the throat of this bright yellow flower; *mimulus* means buffoon. Apparently the two-lobed flower reminds some people of a grinning monkey's face, hence its common name. After the flowers have fallen off, the fused sepals form capsules that contain many seeds. The paired upper leaves clasp the stem. Yellow monkeyflower is native to the western mountains and grows at both lower and higher elevations. Numerous species of *Mimulus*, plants of wet habitats, that often grow in or beside streams and springs, are found in western North America; a few are native to the eastern part of the country. Lewis evidently also collected a pink-flowered *Mimulus*, which Frederick Pursh later painted and named *Mimulus lewisii*. Unfortunately neither specimen has survived the ravages of time.

Several days after collecting the yellow monkeyflower, Lewis and his party came to extensive plains which he called the prairie of the knobs. Here he recorded and collected three new plants.

saw the common small blue flag and peppergrass. The southern wood and two other speceis of shrub are common in the prarie of knobs. perserved specemines of them.

One of the new plants was western blue flag, *Iris missouriensis*, a plant that is poisonous. The other two were the antelope bush and silverberry. He did not collect the southern wood, but earlier in the expedition had collected five of its relatives, all in the family Artemesia, or sagebrush.

Various forms of *Mimulus* are available for the gardener; they may be classified as tender perennials or as annuals. Some are low-growing ground covers and others reach shrubby proportions. The flowers come in shades of red, yellow, apricot, and creamy white. A once popular, highly fragrant species was known as musk flower. Monkeyflower needs a cool situation with moist soil.

BEARBERRY HONEYSUCKLE
Lonicera involucrata

Bearberry Honeysuckle

LONICERA INVOLUCRATA

A COUPLE OF TWIGS OF bearberry honeysuckle, collected on July 7, 1806, are all that exist of this plant in the Lewis and Clark Herbarium. They might have been collected either by Lewis or by Clark. However, because Lewis had the more discerning eye for new and interesting botanical species, it is easy to assume that he is the one to have gathered it. In the remaining months of the expedition, in spite of everyone's eagerness to get home, Lewis still found time to collect new plants and to write at least one more detailed description. The journals do not mention bearberry honeysuckle; Pursh once again provides the important evidence. His annotation on the Herbarium sheet reads "A Shrub within the Rocky mountains found in moist grounds near branches of rivulets. Jul. 7th 1806."

On the same day Lewis and his party were nearing the Missouri River east of present-day Missoula, Montana, at a point where it flows almost due north. They were following rivers that crossed level plains between mountains. The Indians had told them to take this route to the Missouri, which was much shorter than the one they had followed on the way west. They started out early in the morning.

with the road through a level beatifull plain on the North side of the river much timber in the bottoms hills also timbered with pitch pine. no longleafed pine

> *since we left the praries of the knobs. crossed a branch of the creek 8 yds. wid on which we encamped at ¼ m.*

He made similar annotations at various places for the rest of the day, recording the miles from one point to another; he also saw signs of buffalo in the area.

Another plant Lewis collected that day was blanket flower, *Gaillardia aristata,* a garden perennial easily grown from seed. Still another was *Zigadenus elegans,* called death camas because it is poisonous to humans and some animals. Although it does not closely resemble the true camas, its bulb could be mistaken for camas, onion, or a number of other edible bulbs. Fortunately, by now the expedition were finding game to eat and were not desperate for food.

Meanwhile, Clark's party had been heading south. On July 7 they came to a spring of boiling hot "sulferish" water. They put pieces of meat in it, which were cooked in 25 to 32 minutes. Clark did not say whether they then ate the meat. The next day he wrote "The road which we have traveled from travellers rest Creek to this place an excellent road. and with only a few trees being cut out of the way would be an excellent waggon road." This and other remarks in his journal suggest that he was thinking ahead to the time when European-Americans would be heading west.

On July 24, as the group readied themselves to go down the Yellowstone River in their two canoes, they came upon an unoccupied Indian lodge

> *Situated in the Center of a butifull Island thinly Covered with Cotton wood under which the earth which is rich is Covered with Wild rye and a Species of grass resembling the bluegrass and a mixture of Sweet grass which the Indian plat and ware around their necks for its cent which is of a Strong sent like that of the Vinella.*

Clark saw immense numbers of animals that

> *for me to mention or give an estimate of the different Spcies of wild animals on this river particularly Buffalow, Elk Antelopes & Wolves would be increditable. I shall therefore be silent on the Subject further.*

Although Clark observed many plants and animals, his comments were for the most part too general to allow later readers to identify species. His handwriting does not appear on any of the Herbarium sheets. However, there is one plant in the Herbarium that almost certainly was collected by William Clark: *Euphorbia marginata,* snow on the mountain. The annotation, in Pursh's hand, reads that it was collected on July 28 on the Yellowstone, a river that Lewis never traveled. This is a distinctive plant, with a curious flower and pale grayish foliage. The upper leaves are white or white margined. It is a native of the plains and was a new species. When seen in the wild today in the eastern part of the country, it has no doubt escaped from someone's garden.

Bearberry honeysuckle is also called twinberry honeysuckle and black twinberry, because the fruit grow in pairs of purplish black berries. The conspicuous bracts beneath the flower and fruit turn deep red. These bracts and the twin berries, which are not edible for humans, are distinguishing characteristics of the species. It has yellow flowers, rather large leaves, and is found in cool, moist soil. It is a rather scraggly shrub that reaches ten feet in height and is a food source for Gillett's checkerspot butterflies.

Gumbo evening primrose
Oenothera cespitosa

Gumbo Evening Primrose

OENOTHERA CESPITOSA

LEWIS ARRIVED AT THE Missouri River above the Great Falls on July 11, 1806, and remained for several days to rest the horses and arrange for portaging around the falls to be carried out when the rest of the group arrived.

It was probably during Lewis's stay at the Great Falls that he collected the gumbo evening primrose. The fragrant white flower of this native plant can be up to four inches across. It first opens in the evening and is pollinated by night-flying moths. After that it stays open during the day and over time ages from white to pink. Bloom time for the gumbo evening primrose is May to July. The tube of the flower, which is up to three inches long, may appear to be a stem. However, this plant often has no above-ground stem. The scallop-edged leaves and the flowers grow out of the crown of the root. Its long thick tap root and extensive root system help it survive in hot dry climates. This primrose's habitat is on the high plains where the fine silty soil is heavy with clay. This soil type is known as gumbo or gumbo till. Lewis noticed that the soil at the falls had changed from that in the valleys and foothills of the mountains.

> *the land is not fertile, at least far less so, than the plains of the Columbia or those lower down this river, it is a light coloured soil intermixed with a considerable proportion of coarse gravel without sand, when dry it cracks and appears thursty and is very hard, in it's wet state, it is as soft and slipry as so much soft soap. the grass is naturally but short and at present has been rendered much more so by the graizing of the buffaloe, the whole face of the country as far as the eye can reach looks like a well shaved bowlinggreen.*

One senses Lewis's delight with the Great Falls and surrounding land in various passages, such as the one he wrote just after they had arrived: "the air was pleasant and a vast assemblage of little birds which croud to the groves on the river sung most enchantingly." The grandeur of the falls, which had so impressed him the previous summer, now induced him to make several sketches of them. Buffalo covered the plains in numbers that he estimated to be ten thousand within a two-mile area.

One evening one of the expedition's hunters failed to return to camp; he had had to climb a tree to save himself from an approaching bear. During the night he descended and the next morning returned to camp. Hearing of this, Lewis recalled his own experiences with strangely behaving animals the year before and wrote "there seems to be a sertain fatality attatched to the neighbourhood of these falls, for there is always a chapter of accedents prepared for us during our residence at them."

The captain and three other men left the rest of the party on July 17 and headed on horseback for the Marias River, which flows into the Missouri below the falls. Because the Marias is a tributary of the Missouri, Lewis hoped to find its headwaters and thereby determine the northern boundary of the Louisiana Territory.

The only violent encounter with Native Americans during the length of the entire expedition occurred on this side trip. The Nez Perce had warned Lewis and Clark about their enemies, the Blackfeet, who were the dominant tribe on the northern plains. On July 27 the four men engaged in a fight with a small band of Blackfeet, one of whom was killed and another badly wounded. Lewis and his men rode furiously to the Missouri, where they found the rest of their party. They abandoned their horses for five canoes and a pirogue and from that time on traveled by water.

On August 7, where the Yellowstone River flows into the Missouri, instead of meeting Clark they found a note from him attached to a pole. He had gone farther downriver to find more abundant game and escape the mosquitos. One more unfortunate incident befell Lewis before rendezvousing with Clark. He was accidentally shot in the thigh by one of his men, not seriously but enough to cause him much discomfort for a few weeks. On August 12 they met two hunters from Illinois, who had passed Clark just the day before. Lewis's party proceeded on and later that day caught up with Clark and his party.

> *at 1 P.M. I overtook Capt. Clark and party and had the pleasure of finding them all well. as wrighting in my present situation is extreemly painfull to me I shall desist untill I recover and leave to my frind Capt. C. the continuation of our journal. however I must notice a singular Cherry which is found on the Missouri in the bottom lands about the beaverbends and some little distance below the white earth river.*

He described this plant in detail and thus ended his journal.

The reunited expedition continued down the Missouri. The next day they met three traders who gave them some whiskey, the first they'd had in over a year, and that night they stayed up late singing songs. They had replaced the tattered rags of clothing they'd been wearing for some time, so they would look presentable for their arrival in St. Louis. On September 2 they encountered more traders, asked after President Jefferson, and learned about the duel between Alexander Hamilton and Aaron Burr. When they came to small villages they were greeted with much enthusiasm, and word of their return began to spread. They abandoned the two canoes that had been lashed together high on the Yellowstone River. The sight of cows on the riverbank—the four-legged kind, not the roots—filled them with joy. On September 23 they reached the Mississippi River and descended to St. Louis; they fired their guns in salute and received a hearty welcome from the entire town. The Corps of Discovery had returned safely, to the surprise of many, after a two-and-a-half-year 8,000-mile tour across the continent.

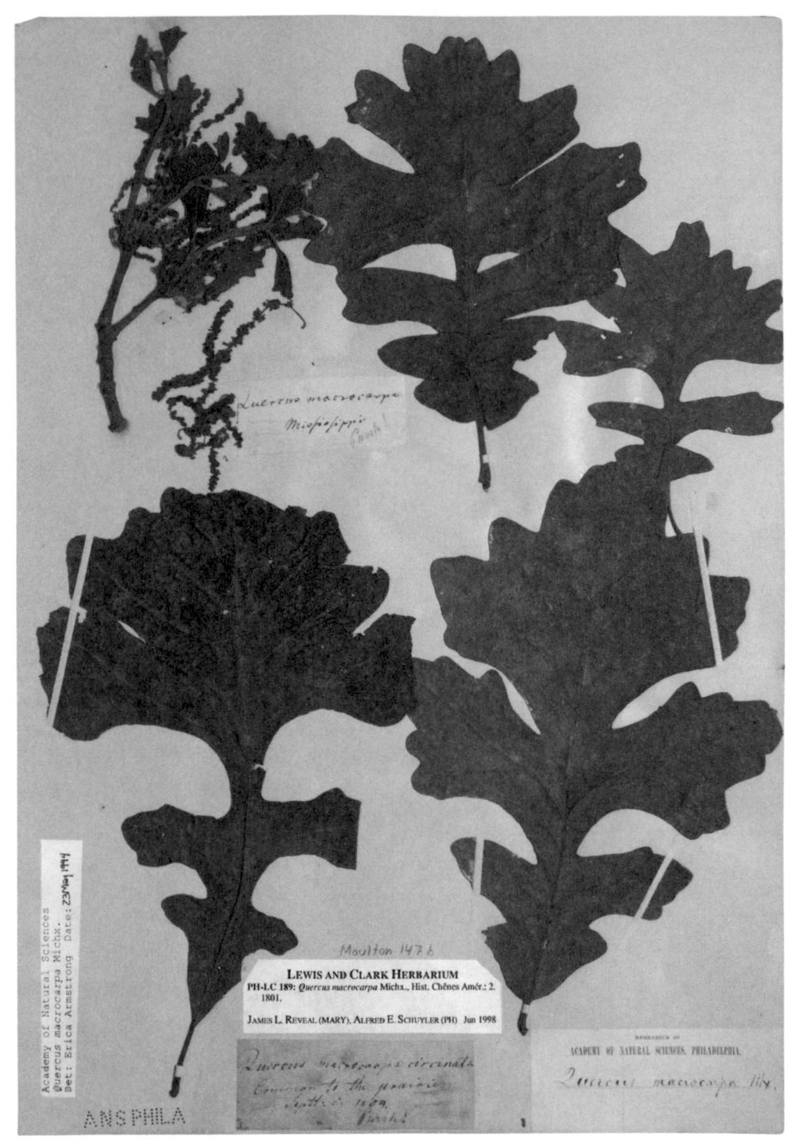

WHEN LEWIS WROTE A DETAILED DESCRIPTION IN THE FALL OF 1805 OF BUR OAK, *QUERCUS MACROCARPA*, THE EXPEDITION WAS ON THE MISSOURI RIVER. THE CATKINS IN THE UPPER RIGHT ARE FROM A LATER COLLECTION, PROBABLY MADE BY THOMAS NUTTALL. THE DEEPLY INDENTED LEAVES ARE TYPICAL OF BUR OAK; THE ROUNDED LOBES ARE TYPICAL OF THE WHITE OAK GROUP, TO WHICH BUR OAK BELONGS.

LEWIS AND CLARK HERBARIUM

The botanical prizes of the Corps of Discovery are housed in the Lewis and Clark Herbarium at the Academy of Natural Sciences in Philadelphia. The 226 pressed and dried specimens represent Lewis's efforts to fulfill President Jefferson's request to bring back plants, "especially those not known in the U.S." Many of them are native to the Rocky Mountains and farther west to the Pacific coast. Some 178 of these plants were newly introduced to Western science. A few more of Lewis's plants are at the Royal Botanic Gardens at Kew in England, and a few collected by later botanists are thought to have been placed on the same specimen sheets with Lewis's plants. The total number of plants collected by Lewis is thought to be somewhere between 232 and 238.

Not all the plants that Lewis collected have been preserved. All those he gathered during the spring and summer of 1805 and stored in a cache at the base of the Rocky Mountains were destroyed by rising springtime waters. Others were lost in transit, were destroyed by insects or mold, or disintegrated over time. Those that survived the boat ride down the Missouri were unloaded in St. Louis and carried overland to Washington, D.C., and to Philadelphia. Some were taken to England and then brought back.

In the spring of 1803, a year before the start of the expedition, Jefferson sent Lewis to Philadelphia to meet with prominent men of science. Many of them were, like Jefferson, members of the American Philosophical Society, which was founded in 1743 and had become the country's premier learned society. From these men Lewis learned about latitude and longitude, celestial navigation and the use of a sextant; he was advised to watch for fossil remains of mastodons and possibly even live mastodons. Because he would be gone for a long time he could not bring back live plants; he had to learn how to press and dry them.

Lewis mostly likely used the methods of other botanists of the day to preserve the plants. The general assumption is that Benjamin Smith Barton showed Lewis what to do. First the botanist should collect a specimen or specimens with as many parts as possible that will determine the plant's identity, including leaves; stems; roots; flowers, including stamens and pistils; hairs and spines; fruit and seed. The time of year determines which plant parts are available. The plant specimens are carefully laid out between two sheets of paper, such as blotting paper or newsprint, that will absorb moisture. The sheets are then placed in a press, which may simply be two pieces of wood, perforated to let moisture escape, and held tightly together with straps. Many specimens can be placed in one press. Once dry, they are removed to an airtight and watertight storage container. There is no mention of plant presses, blotting paper, or other items specifically to be used in preserving plants among the supplies Lewis purchased for the expedition. He did list six brass ink stands and one hundred quills, plus ink and paper.

The collector must also label each specimen, indicating where and when it was found, with a few observations on the habitat. Unfortunately only thirty-four labels in Lewis's handwriting exist. These are written on mauve blotting paper. The plant specimens were all removed at a later date from their original sheets. Many now bear labels in the handwriting of the young German botanist Frederick Pursh, who very likely copied Lewis's notations onto his own labels.

Lewis's first batch of plants, which he sent to the president in the spring of 1805, went to the American Philosophical Society. Barton was expected to examine them but never did. The year after Lewis returned, he took the rest of his plant specimens to Philadelphia. There he handed them over to Frederick Pursh, who at the time was living with Bernard McMahon and doing some plant collecting for Barton. McMahon, the Philadelphia nurseryman, had recommended Pursh as the best person to write formal botanical descriptions of Lewis's plants. After that Lewis left for St. Louis and never again met with Pursh. In 1811 Pursh published his work on North American plants, *Flora Americae Septentrionalis*, extensively based on the plants that Lewis had collected. By that time Lewis was dead and Pursh had left the United States for England, taking with him some of Lewis's plants. He assigned scientific names in Latin to the plants, often honoring Lewis and Clark. *Lewisia rediviva* for bitterroot, *Clarkia pulchella* for ragged robin, and *Linum lewisii* for perennial flax are just a few.

For many years Lewis's plants, except for those that Pursh took to England, remained at the American Philosophical Society, where the journals still reside. The Academy of Natural Sciences was founded in 1812 and by the end of the century was sponsoring scientific explorations worldwide. In the late nineteenth century, interest revived in Lewis and Clark, and the plant collection was transferred to the Academy. It seems nothing short of miraculous that so many of these fragile pieces of plants, neglected for years, survive today.

In addition to pressing specimens, Lewis collected various seeds and roots. From Fort Mandan he sent home to the United States seeds of corn, beans, peas, and tobacco grown by the Arikara, a North Dakota Indian tribe that practiced agriculture rather than depending on hunting. Thomas Jefferson, with his love of vegetables, was much interested in plants grown from these seeds.

Although Lewis was not a trained botanist, he deserves a place in the pantheon of American botany. Many of the great men of science of that time are thought of as naturalists, a term not applicable to highly specialized scientists. John Bartram, so successful as a plant collector in the eighteenth century, was also a farmer, built his own house, which still stands, and experimented with hybridizing plants. Thomas Nuttall, trained as a printer, distinguished himself in ornithology after many years of botanizing. William Clark, military officer and Indian affairs expert, was a superb mapmaker and the first to present an accurate picture of the American west. And Lewis was an astute botanist, ecologist, and zoologist, as well as being a courageous explorer and private secretary to a president, who showed great wisdom when he invited Clark to share the fatigues, dangers, and honor of what turned out to be their greatest enterprise.

SOURCES

Abrams, Le Roy, and Roxana Stinchfield Ferris. *Illustrated Flora of the Pacific States: Washington, Oregon, and California.* 4 vols. Stanford, Cal.: Stanford University Press, 1940–60.

Allan, Mea. *The Hookers of Kew.* London: Michael Joseph, 1967.

Allen, John Logan. *Passage through the Garden: Lewis and Clark and the Image of the American Northwest.* Urbana: University of Illinois Press, 1975.

Barbour, Michael G. et al. *Terrestrial Plant Ecology*, 3rd ed. Menlo Park, Cal.: Benjamin Cummings, 1999.

Barkley, T. M., ed. *Flora of the Great Plains.* Lawrence: University Press of Kansas, 1986.

Barr, Claude A. *Jewels of the Plains.* Minneapolis: University of Minnesota Press, 1983.

Berkeley, Edmund, and Dorothy Smith Berkeley. *The Life and Travels of John Bartram.* Tallahassee: University Presses of Florida, 1982.

Betts, Edwin M., and Hazlehurst Bolton Perkins. *Thomas Jefferson's Flower Garden at Monticello.* Charlottesville: University Press of Virginia, 1986.

Blunt, Wilfrid. *In for a Penny: A Prospect of Kew Gardens.* London: Hamish Hamilton, 1978.

Brown, Deni. *Encyclopedia of Herbs and Their Uses.* New York: Dorling Kindersley, 1995.

Bush-Brown, Louise and James. *America's Garden Book.* New York: Charles Scribner's Sons, 1939.

Clausen, Ruth Rogers, and Nicolas H. Ekstrom. *Perennials for American Gardens.* New York: Random House, 1989.

Craighead, John J. et al. *A Field Guide to Rocky Mountain Wildflowers.* Boston: Houghton Mifflin, 1963.

Cutright, Paul Russell. *Lewis and Clark: Pioneering Naturalists.* Urbana: University of Illinois Press, 1969.

Cutright, Paul Russell. "Well-Traveled Plants of Lewis and Clark," in *We Proceeded On.* Great Falls, Minn.: The Lewis and Clark Trail Heritage Foundation, Feb. 1978.

Elias, Thomas. *The Complete Trees of North America: Field Guide and Natural History.* New York: Times-Mirror, 1980.

Everett, Thomas H. *New York Botanical Garden Illustrated Encyclopedia of Horticulture.* 10 vols. New York: Garland Publishing, 1980–82.

Favretti, Rudy and Joy. *For Every House a Garden: A Guide for Reproducing Period Gardens.* Hanover, N.H.: University Press of New England, 1990.

Fernald, Merritt Lyndon. *Gray's Manual of Botany,* 8th ed. New York: D. Van Nostrand, 1950.

Gabrielson, Ira N. *Western American Alpines.* New York: Macmillan, 1932.

Gleason, Henry A. *The New Britton and Brown Illustrated Flora of the Northeastern United States and Adjacent Canada.* Hafner, N.Y.: New York Botanical Garden, 1968.

Gleason, Henry A., and Arthur Cronquist. *Manual of Vascular Plants of Northeastern United States and Adjacent Canada.* New York: D. Van Nostrand, 1963.

Graustein, Jeannette E. *Thomas Nuttall, Naturalist: Explorations in America, 1801–1841.* Cambridge: Harvard University Press, 1967.

Hatch, Peter J. "Bernard McMahon, Pioneer American Gardener," in *Twinleaf.* Charlottesville, Va.: Thomas Jefferson Foundation, 1993.

Hightshoe, Gary L. *Native Trees, Shrubs, and Vines for Urban & Rural America.* New York: Van Nostrand Reinhold, 1988.

Hitchcock, C. Leo., and Arthur Cronquist. *Flora of the Pacific Northwest: An Illustrated Manual.* Seattle: University of Washington Press, 1973.

Huxley, Anthony et al. *The Royal Horticultural Dictionary of Gardening.* London: Macmillan, 1992.

Jackson, Donald, ed. *Letters of the Lewis and Clark Expedition, with Related Documents, 1783–1854*, 2nd ed. Urbana: University of Illinois Press, 1978.

Leighton, Ann. *American Gardens of the 18th Century*. Boston: Houghton Mifflin, 1976.

Leighton, Ann. *American Gardens of the 19th Century*. Amherst: University of Massachusetts Press, 1987.

Leighton, Ann. *Early American Gardens*. Boston: Houghton Mifflin, 1970.

Kershaw, Linda, Andy MacKinnon, and Jim Pojar. *Plants of the Rocky Mountains*. Edmonton, Alb., Can.: Lone Pine, 1998.

McMahon, Bernard. *McMahon's American Gardener*, 11th ed. New York: Funk & Wagnalls, reprint 1976.

Moerman, Daniel E. *Native American Ethnobotany*. Portland, Ore.: Timber Press, 1998.

Morley, Brian D. *Wild Flowers of the World*. New York: G. P. Putnam's Sons, 1970.

Moulton, Gary E., ed. *The Journals of the Lewis and Clark Expedition*. 13 vols. Lincoln: University of Nebraska Press, 1983–2001.

Munz, Philip A. *California Mountain Flowers*. Berkeley: University of California Press, 1963.

Nelson, Ruth Ashton. *Plants of Rocky Mountain National Park*. Washington, D.C.: U.S. Government Printing Office, 1953.

Niering, William A., and Nancy C. Olmstead. *The Audubon Society Field Guide to North American Wildflowers: Eastern Region*. New York: Alfred A. Knopf, 1979.

Nold, Robert. *Penstemons*. Portland, Ore.: Timber Press, 1999.

Nuttall, Thomas. *A Journal of Travels into the Arkansas Territory During the Year 1819*. Savoie Lottinville, ed. Norman: University of Oklahoma Press, 1980.

Orr, Robert T. and Margaret C. Orr. *Wildflowers of Western America*. New York: Alfred A. Knopf, 1974.

Phillips, Roger, and Martyn Rix. *Perennials*. vols. 1 and 2. New York: Random House, 1991.

Pojar, Jim, and Andy MacKinnon. *Plants of the Pacific Northwest Coast*. Vancouver: Lone Pine, 1994.

Pursh, Frederick. *Flora Americae Septentrionalis*. London: White, Cochrane, 1814.

Reveal, James L. et al. *The Lewis and Clark Collections of Vascular Plants*. Philadelphia: Academy of Natural Sciences of Philadelphia, 1999.

Ronda, James. *Lewis and Clark among the Indians*. Lincoln: University of Nebraska Press, 1984.

Savage, Henry Jr., and Elizabeth J. Savage. *André and François André Michaux*. Charlottesville: University Press of Virginia, 1986.

Schacht, Wilhelm. *Rock Gardens*. New York: Universe Books, 1981.

Seymour, E.L.D., ed. *The Wise Garden Encyclopedia*. New York: Wm. H. Wise, 1951.

Spamer, Earle, and Richard M. McCourt. *The Lewis & Clark Herbarium: Academy of Natural Sciences Digital Imagery Study Set*. Philadelphia: Academy of Natural Sciences of Philadelphia, 2002.

Stephens, H. A., *Woody Plants of the North Central Plains*. Lawrence: University of Kansas Press, 1973.

Time Life Editors and James Underwood Crockett. *Flowering Shrubs*. New York: Time, 1972.

U.S. Department of Agriculture, Forest Service. *Silvics of North America*. Russell M. Burns and Barbara H. Honkala, tech. coord. Washington, D.C.: 1990.

Woodward, Marcus. *Leaves from Gerard's Herball*. Boston: Houghton Mifflin, 1931.

Zucker, Isabel. *Flowering Shrubs and Small Trees*. Derek Fell, rev. New York: Friedman/Fairfax, 1995.

ACKNOWLEDGMENTS

The resources of the Charles E. Shain Library at Connecticut College, the Connecticut Public Libraries, and the Academy of Natural Sciences library were all essential to the preparation of the manuscript. I particularly want to acknowledge Marion Shilstone, Director of Information Resources Team; also Laurie Deredita, Special Collections and Archives Librarian; and the reference library staff. Rick McCourt at the Academy of Natural Sciences showed me plant specimens in the Lewis and Clark Herbarium and, along with Earle Spamer, revealed the importance of this collection. *The Journals of the Lewis & Clark Expedition* edited by Gary E. Moulton and published by the University of Nebraska provided the raw material. Emma Sweeney of Harold Ober Associates played a vital role in making it all happen.

My husband, Neild Oldham, also a writer, knew when to stay out of my way, when to offer much needed words of encouragement, and, of course, when to read and comment on the manuscript.

—Susan H. Munger

Members of the staff of The Academy of Natural Sciences of Philadelphia, Bob Peck, Earle Spamer, and Rick McCourt, were of great help from the inception of this project. Botanists, gardeners, and teachers at the New York Botanical Garden helped me to find specimens and encouraged my artistic development. Members of the Native Plant Society, North Fork Chapter, Missoula, welcomed my many questions and enthusiastically shared their knowledge. Sheila Morrison introduced me to several native plants that she had culitivated in her garden and was a tremendous help identifying species and supplying me with source materials. Kelly Chadwick's love and knowledge of plants was infectious; she gave generously of her time, introducing me to the native plants growing at the University of Montana. Don Hall captured the original watercolors on film.

T. Peter Bennett has been my mentor for the past forty years. His genuine interest in my work has been an inspiration. And my husband, Wayne Thomas, alternately helped me to stay on track and to relax and enjoy the process of painting.

—Charlotte Staub Thomas

ARTIST'S NOTE

Living plants, most growing in gardens and parks, were the models for the paintings in this book.

A magnificent Osage orange grows in Central Park, New York City, and another just outside Chantecleer Garden in Pennsylvania. Bur oak, bearberry, snowberry, and Oregon grape holly were at the New York Botanical Garden. I was able to grow camas from a bulb on my window sill in New York City.

I found bear grass, glacier lily, and bearberry honeysuckle at Glacier National Park. The remainder of the plants grew in Missoula, Montana, some at the Native Plant Garden at the University of Montana, others in the fields and parks around town. Wood lily and gumbo evening primrose were in Sheila Morrison's garden.